P9-EKS-690

Library of Industrial and Commercial Education and Training
ADVISORY EDITORS:
B. H. Henson, B.Sc.(Econ.) T. F. West, D.Sc., Ph.D., F.R.I.C., A.M.I.Chem.E.

CHEMICAL INDUSTRY DIVISION
General Editor: W. Sabel

CHEMICAL PLANT AND ITS OPERATION

CHEMICAL PLANT AND ITS OPERATION

BY

T. M. COOK, M.Sc., Ph.D., C.Eng., M.I.Prod.E., F.R.I.C.

AND

D. J. CULLEN, C.Eng., M.I.Mech.E.

Thomas Morson & Son Limited, a subsidiary of
Merck Sharp & Dohme International,
Enfield, Middlesex

PERGAMON PRESS

PERGAMON PRESS LTD.

OXFORD · LONDON · EDINBURGH

NEW YORK · TORONTO · SYDNEY

First edition 1969

Library of Congress Catalog Card No. 73–88305

Printed in Great Britain by A. Wheaton and Co., Exeter

08 006527 9

Contents

Preface

THIS book has been written as a textbook for the chemical operator who is interested in developing his skill and gaining greater job satisfaction. The development of the modern chemical industry depends on the availability of increasing numbers of trained operators who can take advantage of the new operating techniques and the wide range of plant and equipment now available to increase productivity, reduce manufacturing costs, and improve standards of safety.

Throughout the book descriptions of chemical plant and operations have been given from a practical standpoint and the theoretical principles involved have been kept to the minimum necessary. Particular emphasis has been placed on the technique and safety aspects of operations.

The scope of this book is designed to meet the needs of chemical operatives who are preparing for the examinations for the ordinary and advanced certificates in chemical plant operation, and for those taking chemical technician courses. It will also be of value to all those involved in chemical manufacture, particularly supervisors and maintenance staff.

We are indebted to the Editor of this series of books, Mr. W. Sabel, for his helpful advice and criticism. We are also very grateful to Mr. P. A. J. Cripps, who prepared the diagrams, and to Mr. C. H. G. Cook, Mr. H. J. Whitehouse, and Mr. G. O. Young for their assistance in the preparation of the manuscript. Thanks are also due to our colleagues who have helped in so many ways.

T. M. COOK
London D. J. CULLEN

Publisher's Foreword

THE Industrial Training Act has resulted in an increase in the number of people now being trained or re-trained. LICET books are intended to provide suitable texts which will be easy to read and assimilate for those employed in industry and commerce who are receiving further education and training as a part of their employment. It is hoped that they will be particularly suitable for those attending courses leading to the examinations of the City and Guilds of London Institute, the Regional Examining Unions and other examining bodies.

The books are essentially straightforward, simple and practical in their approach and are designed to provide all the basic knowledge required for a particular trade or occupation. They are structured in such a way that the subject is broken down into convenient and progressive components, and are written by authors specially chosen for their expert knowledge and for their practical and teaching experience of their subjects.

Where appropriate, emphasis has been placed on safety training. In some subjects separate manuals on safety and safety training will be provided; in other texts, authors have been encouraged to emphasise safety precautions at relevant points, or to devote a separate chapter to these matters.

LICET books are published in a number of subject divisions, with each division controlled by a specialist editor responsible for selecting authors and providing guidance and advice to both authors and publisher. It is hoped that the series will make an important contribution to further education and industrial training.

ROBERT MAXWELL
Publisher

CHAPTER 1

Materials of Construction and Corrosion

THIS chapter describes, in general terms, the materials—metals and non-metals—used in the construction of chemical plant, and discusses the main factors which determine their selection. The major consideration of the suitability of a material is its corrosion under the particular service conditions required. Other factors determining the choice of a material include its availability, mechanical properties, physical and chemical properties, heat resistance and cost of fabrication, installation and maintenance. Building materials are not mentioned specifically, but many of these are used in the construction and installation of chemical plant.

CORROSION AND ITS PREVENTION

There is no simple explanation of corrosion, but it can be defined broadly as chemical attack on metals or alloys under certain specified conditions, causing thinning, cracking, pitting, blistering, or erosion. These conditions arise from a combination of a number of variable factors, such as temperature, humidity, exposure, stress, abrasion, pressure or vacuum, or contact with certain chemicals. For example the rusting of iron—which is the most common form of corrosion—is caused by its exposure to air and water, but iron will not rust in air which is kept absolutely dry.

Corrosion of chemical plant must be considered in two ways—internally and externally. If a corrosive chemical is contained in a vessel, or transmitted through a pipeline or pump then, ideally, the material of construction must not be attacked by that chemical under any conditions. On the other hand, any item of equipment or plant and the building structure in which it is placed must be resistant to the surrounding atmosphere.

It is common practice to protect structural steel and timber in any atmosphere by the application of paint to the clean surface, to prevent rusting or the penetration of damp. In chemical factories where there is a risk of more severe atmospheric corrosion and spillage of corrosive chemicals, the same principle must be used; namely the application of a skin of resistant material to equipment and buildings. Paints used for this purpose range from the conventional oil-based type, through special acid- or alkali-resistant types to those based on chlorinated rubber and plastics. Plastic paints use materials such as acrylic, epoxy, and polyurethane resins. The range of special resistant paints is too wide to discuss here, but as a general guide they are produced to resist acidic or alkaline conditions, moisture, abrasion, extremes of temperature, or solvents. In more severe conditions metals may be protected against corrosion by coating with plastics such as PVC, or reinforced plastics such as polyester—glass fibre.

Whatever protection is applied to any material, care must be taken to avoid damaging the paint or plastic layer. Full protection is often gained by applying several coats of different materials, therefore even surface damage can lead to a total breakdown of the protective surface. In known areas of spillage, special protection may be provided by drip-trays or aprons constructed in suitably resistant materials.

SELECTION OF MATERIALS

Materials of construction are selected to give adequate service at lowest cost. The main considerations are the effect of the chemical on the selected material and the effect of the material on the chemical.

If only one chemical is being handled in the vessel or pipeline, it is often possible to select a material that will be completely resistant. Should this selected material be very expensive or not readily available it may be practical to choose one which has a rate of attack sufficiently slow to make replacement an economical proposition. If this is done the effect of that corrosion on the chemical being handled and the final product must be checked.

When several different chemicals are to be handled it may be necessary to run pipes of different materials of construction along the same route and have charge tanks of similar sizes but dissimilar materials of construction feeding the same reaction vessel. In the reaction vessel, however, this separation cannot be maintained and the selection of a material completely resistant to (1) all the chemicals to be fed into the vessel and (2) all the chemicals resulting from the reaction that takes place in the vessel, may be impossible. There are, however, several materials that are resistant to a wide range of chemicals and it is mainly in the construction of reaction vessels that these materials are used.

Consideration is now given to some of the many materials of construction used in chemical plant manufacture and their characteristic properties. The more widely used materials are classified as follows:

Classification of Materials of Construction

1. *Metals*
   (a) *Ferrous metals.* Cast iron, iron alloys, mild steel, steel alloys (for example stainless steels).
   (b) *Non-ferrous metals.* Aluminium, tin, nickel, chromium, copper, zinc, lead, non-ferrous alloys (for example bronze, brass).

2. *Non-metals*
   (a) *Naturally occurring materials.* Wood, stone, natural rubber.
   (b) *Manufactured materials.* Synthetic rubber, cement, stoneware, graphite, glass, plastics (thermoplastic and thermosetting).

FERROUS METALS

Cast Iron

Cast iron is produced by pouring molten, impure "pig" iron into moulds. It is a common material of construction and although relatively brittle, the hard skin formed in the casting by the chilling of molten metal against the cold mould is more resistant to acid gases than mild steel. However, it must be remembered that those areas of casting that have been machined in the course of manufacture are much less resistant because the hard outer skin has been removed. Castings for electric motors, motor starters, and gear boxes are usually manufactured in cast iron and it is convenient to use standard materials which have at least a reasonable life in corrosive atmospheres.

Cast iron is not readily fabricated and requires the construction of a pattern and a mould making the casting of individual items very expensive. It is, however, used throughout the chemical industry as a base material for lining with rubber or plastic in the manufacture of pump and valve casings and some glass-lined vessels. It is usually confined to standard items that are widely used.

Iron Alloys

The most usual alloys of iron used in chemical industry are those of silicon and iron, containing approximately 14–15 per cent silicon which are resistant to concentrated mineral acids, for example sulphuric acid. It is commonly used in the manufacture of pumps and pipelines but unfortunately it is a brittle material that can be broken very easily, and it has to be handled carefully; it is not readily weldable.

Mild Steel

Mild steel is a mixture of iron and carbon and differs from harder steels in that its carbon content is relatively low (0·1-0·25 per cent). It is probably the most common material of con struction in the chemical industry but it is the most widely attacked of all materials. It is attacked by most acids, caustic solutions will cause embrittlement at high temperature, and in normal atmospheric conditions rusting takes place. Although it is not attacked by organic solvents, it is essential that moisture is excluded from equipment.

It is, however, a cheap and readily available material which can be fabricated easily by well-established techniques. It is used as a base material for lining with rubber, glass, or plastic, or it may be coated internally with a plastic or a paint. In other instances it is used to support a thin layer of expensive material. In most cases it must be treated to prevent corrosion from the surrounding atmosphere.

Cast Steel

Cast steel has a similar chemical resistance to that of cast iron but is less brittle and is usually substituted for the latter where its better physical properties are required.

Steel Alloys

The most common of the steel alloys are usually referred to as stainless steels and cover a very wide range of alloys containing, in the main, chromium, nickel, and silicon. Stainless steels have a very wide range of application outside as well as inside the chemical industry. The alloys containing 12–18 per cent of chromium are generally resistant to corrosion; the lower chromium content alloys are often used for reasons of cleanliness in food manufacture, but at the higher chromium level thay are resistant to many acids and acid gases. These stainless steels can be fabricated satisfactorily but care often has to be taken to avoid corrosion taking place at areas of high stress (where the metal has been formed) and at welds. The addition of small quantities of molybdenum and/or titanium and/or niobium can have a marked effect on the resistance to certain acids and in some cases make fabrication easier. The proportion of chromium to nickel in the alloy is critical and is usually specified. For example, 18/8 Stainless Steels—having 18 per cent chromium and 8 per cent nickel content.

Other stainless steels containing 10 per cent nickel and up to 20 per cent chromium are used in structural work because of their superior mechanical properties.

NON-FERROUS METALS

A wide range of non-ferrous metals is used in the construction of chemical plant and some of those most commonly used are now discussed.

Aluminium

Aluminium is a silver-white metal which is very light; it is a good conductor of heat and electricity, malleable (easily beaten or rolled into sheets), and ductile (readily stretched) and of a high tensile strength. The metal slowly oxidizes in moist air, a thin protective film of oxide being formed. It is attacked by dilute hydrochloric acid, concentrated sulphuric acid, sodium hydroxide, potassium hydroxide, and brine, but is almost unattacked by nitric acid in all concentrations. Its alloys with copper are called aluminium bronzes. Its many uses as a material of construction, as the metal or its alloys, include hoppers, drums, ventilation hoods, and ducting, utensils, and electric conductors.

Chromium

Chromium is a white-grey, lustrous, hard metal, not tarnished when exposed to air. It is attacked by dilute sulphuric and hydrochloric acid and alkalis, but is resistant to nitric acid, Rarely used by itself, it is used as constituent of stainless steels and chrome-steel or for chromium plating to give a decorative finish.

Copper

Copper is a red metal, very malleable and ductile and a good conductor of heat and electricity. It tarnishes when exposed to air, slowly forming a green film of basic copper salts called *verdigris*. Copper is easily fabricated, but care must be taken as it tends to work-harden; repeated bending of the metal causes it to become brittle. It is used in the construction of heat exchangers, electrical components and cables, and as a constituent of brass, bronze, and other alloys.

Lead

Lead is a heavy, bluish-grey metal that is very soft, tough, and malleable and can be welded as well as soldered. It is used in sheet form either as a lining or suitably supported in a frame for the construction of tanks, vessels, and pipes. Lead rapidly tarnishes in air and slowly forms a protective coating. It is resistant to many dilute acids at low temperature and is widely used for handling dilute sulphuric acid.

Nickel

Nickel is a hard, grey-white metal, malleable, and resistant to corrosion when exposed to air. Slowly attacked by dilute hydrochloric and sulphuric acid and readily attacked by all concentrations of nitric acid. It is particularly resistant to concentrated sodium and potassium hydroxide and is therefore often used for vessels which contain them. There are several commercially produced grades of nickel which, although expensive, may be fabricated easily: they are used in the construction of pumps, pipes, and fittings, valves, heating coils, drums, tanks, and reactors, and components for glass-lined vessels. It is also used extensively for nickel plating and in alloys, particularly Monel metal and stainless steels.

Tin

Tin is a soft, silver-grey metal, very malleable and ductile but becoming brittle when heated to 200°C or more. Slowly attacked by dilute acids and more readily by concentrated hydrochloric acid and hot sodium hydroxide, its main uses are for containers, tin-plating, and in alloys.

Zinc

Zinc is a white metal, fairly hard and brittle, becoming very brittle when heated above 200° C. When exposed to moist air it forms a thin, white, protective layer of basic salts; for this reason it is applied to iron by a hot dip process, called *galvanizing*. Zinc is attacked by dilute acids (more slowly when the metal is pure) and alkalis. It is rarely used alone as a material of construction, but is used for galvanizing, as a constituent of alloys such as brass and bronze, and for building materials.

Other Metals

Tantalum, titanium, zirconium, and the noble metals—silver, gold, and platinum—all have a very wide resistance to corrosion and are used where "universal" resistance is essential. For example tantalum is widely used to repair damage to glass-lined vessels, for thermometer pockets, and as a lining for certain reaction vessels. All these very desirable metals are very expensive and this precludes their general use.

NON-FERROUS ALLOYS

Brass and Bronze

Both are alloys of copper; brass is an alloy of copper and zinc and bronze is an alloy of copper with tin or other metals, such as aluminium. Brass is a hard alloy which is readily cast and machined. There are two types; that containing 30 per cent zinc is used commercially in sanitary fittings and is usually chromium plated. The group of brasses containing between 40 per cent and 45 per cent zinc have a higher tensile strength, are harder, and have a better resistance to corrosion. These are used widely in marine engineering because of their resistance to sea-water.

There is a wide range of bronzes most of which are used as bearing metals. The most chemically resistant type is aluminium bronze, a light alloy containing 90 per cent copper and 10 per cent aluminium of good mechanical strength. It is mainly used outside the chemical industry for fabricating components for marine purposes because of its resistance to salt water.

Other non-ferrous alloys are also used in the chemical industry and mention must be made of alloys of the "noble" metals. Alloys of gold and platinum, for example, have a high resistance to chlorosulphonation reactions, but are only used very sparingly because of their cost.

There are many other alloys available for particular applications, but unless there is a sizeable demand for them their cost is usually very high. An alloy could be found to resist any particular chemical in given circumstances, but the working conditions must be carefully specified. It cannot be assumed that because a particular alloy (or metal) resists a concentrated acid at a certain temperature, it would not be attacked by that acid when diluted or at a different temperature.

NON-METALS

These materials may be divided into two classes: those which occur naturally, such as wood, stone, and raw rubber, and those which are produced synthetically, such as glass, graphite, stoneware, plastics, and synthetic rubber. Though metals are of prime importance in the construction of chemical plant, non-metals possess certain qualities, which, in spite of their low thermal conductivity and physical strength, make them a preferable material. For example the transparency of glass or the resilience of rubber are often important.

Wood

Wood is a universally available material and is used as building material in roof structures, window frames, doors, and flooring. Broadly there are two types of wood: soft woods, such as deal and pine, and hard woods such as oak and teak. The resistance of wood to chemical attack is generally the same as its resistance to water absorption. The hard woods are therefore most commonly used as they are often very resistant to acids. The main uses in the chemical industry are for acid-resistant flooring, the construction of tanks or vats, agitators, and filter press plates and frames. When wet, all wood will expand and this is often used to advantage in keeping a tank free from leaks. Care must be taken to avoid drying out as warpage is most likely to occur, therefore wooden filter presses and vats are kept damp when not in use.

Stone

Stone is rarely used these days, except as a flooring material and and in some ball mills. It has generally been superseded by manufactured materials in the ceramic group.

Rubber

Rubber, in its soft natural state is a polymer of isoprene: it is only of limited use and for most purposes it must be vulcanized. This process involves the chemical reaction between natural rubber and varying amounts of sulphur under heat treatment. Soft rubbers may contain up to 5 per cent sulphur, whilst at the other end of the range, *ebonites*, which are hard and quite brittle, may contain up to 40 per cent sulphur.

Rubber is chemically resistant to acids (especially dilute solutions) and dilute alkalis, resistant to erosion and abrasion, and a good electrical insulator. However, it is attacked by oxidizing agents and hot alkalis, and organic solvents may cause it to swell.

Soft rubber is used for cable insulation, flexible hose, shock absorbers, as a gasket material, and as a lining for iron and steel vessels and pipelines. When used as a lining, rubber is applied to the vessel with a special adhesive. Hard ebonites may be used for pump or valve linings and, without metal support, for smaller vessels.

Many applications are being found for synthetic rubbers, which are synthetic polymers possessing rubber-like properties. Among those available commercially are butadiene-styrene and butadiene-acrylonitrile (called Buna rubbers), polyisoprene, and polybutadiene. Their properties may be modified considerably more than vulcanized rubbers, particularly with respect to resistance to oxidizing agents, solvents, and oils. Their adhesion to metals, however, is generally poorer.

CERAMICS

A ceramic is defined as a material manufactured from a clay or similar substance and covers a wide range of cements, stoneware, earthenware, glass, porcelain, and silica. Ceramics are hard materials which are resistant to wear and chemical attack, but they are brittle and susceptible to sudden changes in temperature.

Cement

Cement is manufactured by heating a mixture of limestone and clay and grinding the product. The most common form is called Portland cement. It is widely used when mixed with aggregate to form concrete for foundations, tanks, and floors. It is often reinforced by a framework of steel, particularly when used in building structures. Rapid-hardening cement is made from a clay containing more than 30 per cent alumina and has the advantage of maturing the concrete in approximately one twenty-fifth of the time of Portland cement. It is more expensive but can be used almost immediately it has dried. Concrete made with some rapid-hardening cements is resistant to many dilute solutions of acids.

Concrete is not usually used as an acid-resistant material but as a building material it is often superior to steel in resisting acidic fumes. In many cases a decorative finish is more easily maintained on concrete than steel. Concrete tanks and floors are often lined with acid-resistant bricks or tiles or plastic materials.

Another use of cement is that of jointing and bedding material for tiles and bricks of all types, and cement in this context includes many synthetic materials which are produced to meet particular conditions. For example sodium silicate and potassium silicate cements are acid resisting, whereas rubber latex, furan resin, and epoxy resin cements are both acid and alkali resisting. In most cases the suitability of a particular cement for a given condition should be carefully checked with the manufacturer before it is used.

Stoneware

Salt-glazed *earthenware* is manufactured from a wide range of clays which are glazed by adding common salt into the kiln whilst the article is being fired. This produces a sodium aluminium silicate surface which has a good resistance to chemical attack on the otherwise porous earthenware. Salt-glazed pipes are almost universally used for sewage and effluent pipework. Earthenware vessels are also glazed by applying a lead compound which fuses to form a lead silicate. There are several grades of earthenware, stoneware, and chemical stoneware; the latter grades being resistant to most corrosive agents except hydrofluoric acid and hot, concentrated alkalis.

Storage tanks up to 100 gal capacity, vacuum filters, towers up to 42 in. diameter, pumps, valves, pipes, and tiles are manufactured in this material. It is fragile, has a limited resistance to thermal shock and therefore needs to be handled and used carefully.

Clay of high purity produces *porcelain* which is used in the food industry and in laboratories because it is easily cleaned. In the unglazed state, earthenware is used as a filter medium and as a diffuser.

A further development of ceramic materials are those produced from metallic oxides. Aluminium oxide, for example, is a hard material with good mechanical strength and capable of withstanding high temperature, abrasion, and chemical attack. It is widely used for mechanical seal faces, spray jets, pumps, and impellers, and as a milling medium.

Graphite

Graphite is a crystalline form of carbon and in its normal state is slightly porous. It is usually rendered impervious by impregnation with a selected phenolic resin or with carbon, the latter producing a truly all-carbon graphite. It is not only resistant to most chemicals except strong oxidizing agents, it is light, is resistant to thermal shock, and has a high thermal conductivity. It is, therefore, an ideal material of construction for heat exchangers handling corrosive liquors or vapours but it has a low coefficient of linear expansion and this makes it difficult to install in conventional tube exchanges. It is available in block form.

The block is fabricated with rows of holes arranged so that alternate planes are at 90 degrees to each other, process liquor passing in one direction and the service liquor passing in the other direction. This results in a light, compact, and efficient heat exchanger with wide application in the chemical industry. Graphite is also used in the manufacture of pumps, pipes, valves, and blocks or bricks for the lining of furnaces.

Glass

Glass is produced by fusing the oxides of silicon, boron, or phosphorous with a basic oxide or sodium, calcium, magnesium, or potassium, and cooling the product rapidly to prevent crystallization. It is a transparent, hard, brittle and amorphous material resistant to all acids except hydrofluoric acid.

The cheapest form is *soda-glass*, which is brittle and sensitive to thermal shock, but *borosilicate glass*, for example Pyrex or Hysil, is stronger and has a greater resistance to thermal shock. It has a wide range of application as pipes, valves, heat exchangers, and small vessels. It enables several acids to be handled in the same pipeline and has the added advantage that, since it is transparent, the interior of such equipment is clearly visible.

Glass-linings. To overcome the poor physical strength of glass, whilst utilizing its corrosion resistance, glass-lined iron or steel is extensively used in reaction vessels, pipes, and valves. In the manufacture of glass-lined vessels (for example "Pfaudler") several coats of a thin film (a few thousandths of an inch) of glass are applied to the metal. The composition of the glass varies according to the particular service required. The initial coating will be selected for its adhesive properties and the outer coatings for their resistance to concentrated acids and alkalis, or for their high resistance to thermal shock.

Glass-lined vessels require particular care and may be damaged by abrasion, or blows by hard objects, or sudden changes in temperature. Even a small flaw in the lining may lead to serious damage when the vessel contains acids, since the acid may seep between the glass and metal casing, causing areas of glass to fracture by the release of gas (formed by a chemical reaction between the metal and the acid) beneath the surface.

PLASTICS

The use of plastics as materials of construction has expanded rapidly in recent years and now they compete with metals, glass, and wood in many applications (see Table 1). The chemical structure of plastics is complicated, but most of them are polymers, that is they consist of large numbers of small, identical groups of atoms joined together to form one chain-like molecule.

There are two main types of plastic materials, namely thermoplastics and thermosetting resins (or *thermosets* as they are often called). Thermoplastics soften on heating so that the original moulded shape may be altered repeatedly and are weldable. This lack of heat resistance may, however, be a disadvantage in use. Thermosetting resins, once subjected to heat and/or pressure and moulded into a particular shape, permanently set into a solid shape which cannot be altered.

Nowadays there are so many plastics available it is only possible to mention a few of the more important types to indicate their range and application.

Phenol-formaldehyde (*phenolic resins*). A range of rather brittle, dark brown or black thermosetting resins. The original material was

TABLE 1. COMPARISON OF PROPERTIES OF METALS AND PLASTICS

Metals	Plastics
(i) Pure substances consisting of atoms of a single element	Chains of identical groups of atoms based on carbon with other elements, such as hydrogen and oxygen
(ii) Generally grey or silver colour	Colourless, when pure
(iii) Not transparent	May be transparent. For example, polystyrene, polycarbonates
(iv) Tend to be attacked by mineral acids	Resistant to mineral acids
(v) Not attacked by organic solvents	Attacked by organic solvents, such as carbon tetrachloride, causing swelling
(vi) High melting points	Low melting points
(vii) High density	Low density—about the same as water
(viii) Good conductors of heat and electricity	Poor conductors of heat and electricity

given the trade name Bakelite: the first synthetic resin produced, it was used for electrical fittings and casings. Later variations, less brittle, are used for pump impellers, dip-pipes, agitators, and bottle closures.

Polyethylene (*polythene*). A thermoplastic with the trade name Alkathene which is a polymer of ethylene. It is produced as low-density polyethylene at very high pressure. Its toughness, flexibility, and resistance to chemical attack has given rise to numerous applications such as pipelines, casings, hose, carboys, drum and tank liners, hand tools, buckets, cable covering, and as a packaging material. More recently, a rigid, high-density polyethylene produced at lower pressures is finding increasing application such as tanks and drums.

Polystyrene. A polymer of styrene produced by a variety of methods in many grades. It is a transparent, colourless, rigid material which is rather brittle and attacked by many organic solvents. It is used extensively in packaging, for bottle closures, and for casings. Expanded polystyrene is used as an insulating material, for cushioning glass containers, and for ceiling tiles.

Polytetrafluoroethylene (*PTFE*). A fluorocarbon resin with the trade names Teflon or Fluon. It is a white, waxy solid with some remarkable properties; virtual immunity to chemical attack, resistance to high temperature, and outstanding non-stick properties and smoothness. Its uses include gaskets, seals, diaphragms, gland packings, filters, bearings, pump-linings, hose, and small pumps.

Polyvinyl chloride (*PVC*). A non-flammable material with good chemical resistance produced in flexible, semi-rigid, and rigid forms. It is used extensively for protective clothing, hose, electrical cable insulation, as a packaging material, and for drum liners.

Epoxy resins. Thermosetting materials which may be syrupy liquids or solids. Possessing excellent heat and chemical resistance, they are used as adhesives and for paints.

Nowadays, plastics are frequently combined with metals, glass, wood, or other materials. For example polyester and epoxy resins may be reinforced with glass fibre, or phenolic resins built up with asbestos to form laminated structures giving extra toughness and rigidity. Carbon fibres impregnated with resin are now commercially available. They were developed originally for aircraft and spacecraft and are of high strength and stiffness with minimum weight. They are likely to be used in chemical plant in the future.

CHAPTER 2

Materials Handling

IN CHEMICAL factories the storage and conveyance of solids, liquids, and gases—from raw materials to finished products—is essential to safe, economic, and uniform production. Materials handling is an essential feature of all basic unit operations (see Chapter 4). Manual labour is rapidly being replaced by machinery and instruments which provide safer automatic control and operation. Whatever material is handled, by whatever means, it is essential that the chemical operator knows what he is handling, wears the appropriate safety clothing, and always respects chemicals as being potentially hazardous.

HANDLING SOLIDS

Storage of Solids

The simplest form of storage is in heaps in open or covered yards, but this method is not suitable for corrosive, flammable, or explosive materials.

Large bulk containers are usually square or rectangular in shape and constructed in steel, aluminium, wood or concrete. Unless the container is emptied manually, it may be fitted with a tilting mechanism or its lowest part forms a conical or pyramidal hopper. Hoppers may incorporate shaker devices, filters, and special types of valves for discharging lumpy or moist material and multiple discharge outlets.

Ore, or similar inert materials, usually stored in bulk in bins or bunkers filled by conveyor, elevator, crane and grab, or mobile shovel. In many instances the material is removed from these storage bins by similar methods, often being transferred over long distances by conveyors or overhead ropeways.

Mixtures of various raw materials in accurate proportions are often achieved by feeding the constituents from several hoppers at controlled rates onto a common conveyor (see Fig. 2.1).

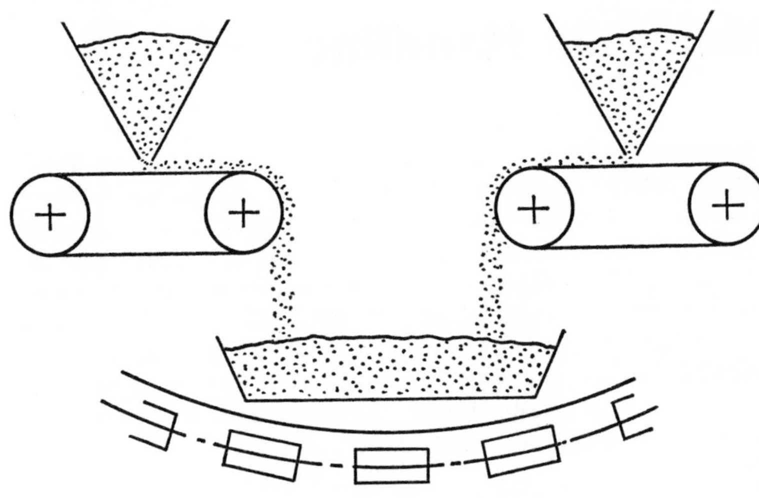

FIG. 2.1. A common conveyor.

Conventional containers for solids, which are manually handled, may be steel, aluminium, plastic, or fibre drums, cartons, barrels, or sacks. Many of these are fitted with an inner lining of rigid plastic or a polythene bag which is resistant to corrosion and readily sealed.

Special regulations exist for the storage of hazardous materials in chemical factories; they specify the type of container; the space between buildings and their construction; the ventilation and the availability of safety and fire-fighting equipment. All containers must be labelled to show their contents clearly—an unlabelled container must not be used under any circumstances.

Conveyance of Solids

The type of conveyor used to transport solids depends on:

(i) Physical state. For example whether the material is a fine powder or lumpy, moist or dry, heavy or light.

(ii) Quantity. The material may be supplied on a continuous or batch basis.

(iii) Hazardous properties. For example whether the solid is explosive, flammable, or poisonous.

(iv) Direction. Vertically and horizontally.

(v) Distance. Between factories or within the factory or plant.

(vi) Containers. The material may be loose or packaged in barrels, cartons, drums, or sacks.

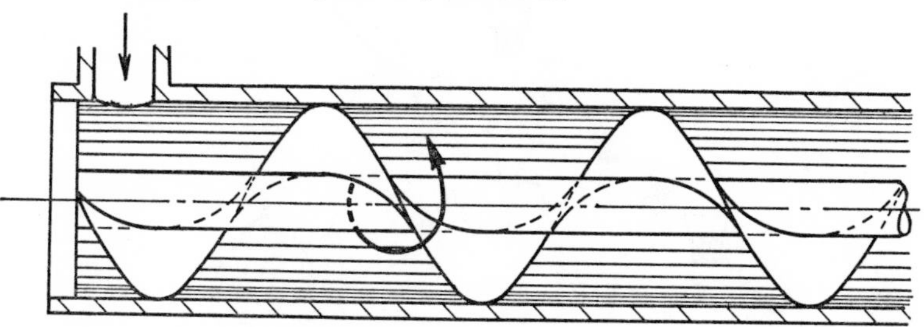

FIG. 2.2. Screw conveyor.

Conveyors

Screw conveyors have been used for many years. They consist of rotating screws which push the material along pipes, shafts, or troughs. The speed of rotation of the screw determines the rate at which the solid is discharged at the end of the conveyor and gives precise control for feeding directly to a reaction vessel (Fig. 2.2).

Belt conveyors are very versatile and the most widely used. They can convey loose or packaged loads over short or long distances, horizontally, or up and down gradual slopes. The endless belts, which are either flat or curved, run over rollers and between pulleys. The drive pulley usually is placed at the delivery or highest end to keep the belt taut. The materials most commonly used for conveyor belts are rubber, plastic, or leather.

A *bucket elevator* is a type of belt conveyor which has buckets fitted to the belt at regular, closely spaced intervals. They are used for lifting solids or liquids vertically, or nearly vertically, and are often filled at the lower end from a horizontal belt conveyor (Fig. 2.3).

Air conveyors or pneumatic conveyors have been developed considerably in recent years and are most suitable for finely divided

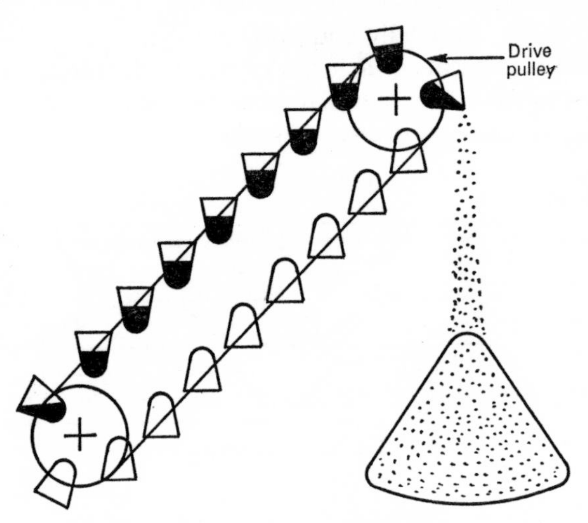

FIG. 2.3. Bucket elevator.

or light materials. An air system, produced by blowing at one end of a pipe, and sometimes sucking at the discharge end or both, entrains the solid particles and carries them along the pipe.

Trucks. Industrial trucks are the most commonly used form of transportation of materials. They are particularly suitable for chemicals packed in containers and are relatively cheap and flexible in their use. There are two types: hand trucks and powered trucks. Hand trucks have been designed to suit the nature of the load to be carried. The main types are sack trucks, drum or barrel trucks, cylinder trucks, and carboy trucks which run on two wheels (Fig. 2.4). Only the correct truck may be used in safety since each type is specifically designed with a lifting platform suitable for the shape of the containers to be carried. Hooks or similar devices are often used to hold the load firmly to the truck.

Platform trucks are load-carrying platforms mounted on four wheels or castors. The platforms, which are generally loaded by hand, are often fitted with some form of superstructure to hold

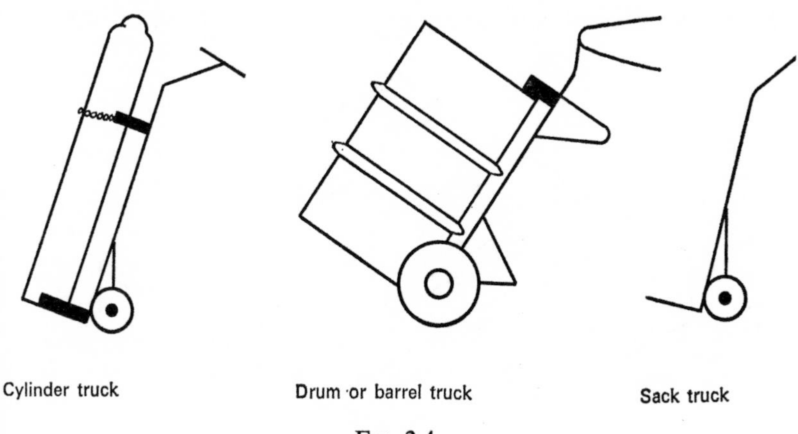

FIG. 2.4.

the load in position. Manoeuvrability is increased if swivel wheels or castors are fitted at one end.

The most versatile power truck is the fork-lift truck which may be powered by an internal combustion engine or batteries. The basic unit is equipped with two forks attached to uprights for lifting standard pallets, but a variety of special-purpose arrangements may be fitted, such as drum grabs and pushing devices. The use of pallets, which are load boards for the assembly of a number of containers to form a "unit load", is now almost universal in chemical factories. They are constructed in wood, steel, or light alloy. The most common type is the double-decked pallet which is illustrated in Fig. 2.5.

Some pallets are fitted with a superstructure which may be

collapsible to conserve storage space of empty pallets. Loaded pallets may be stacked one above the other to make the best use of storage space.

Whatever type of truck is used it is important to remember that it is dangerous to exceed the safe working load specified by the manufacturer.

A recent development is the use of hover pallets which float on a cushion of air on the hovercraft principle. Designed to carry a 1 ton load they present interesting prospects in their future use.

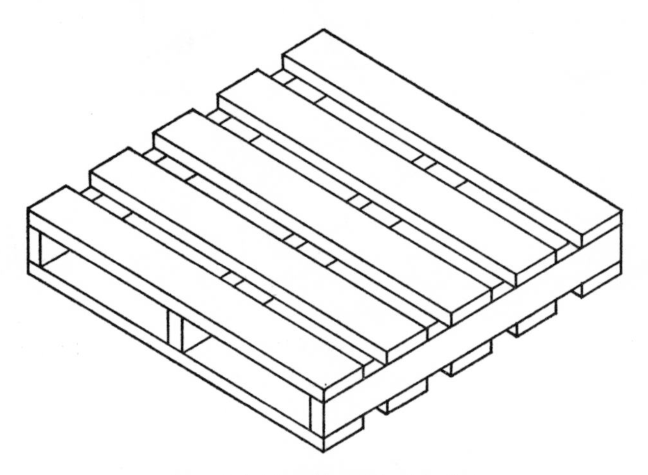

FIG. 2.5. Double-decked pallet.

HANDLING LIQUIDS

Storage of Liquids

Large tanks for the storage of several thousand gallons of liquid may be cylindrical with dished ends or rectangular in shape and mounted vertically or horizontally. They are usually placed in a remote position, mounted on concrete foundations with a surrounding concrete (bund) wall, or placed underground. Materials for tank construction may be stainless steel, aluminium, or mild steel with a rubber or plastic lining for use with corrosive liquids. Tanks are designed for liquid storage at atmospheric, low, or high pressure. Some liquids, such as water, may be stored at atmospheric pressure in open tanks, but most tanks are usually closed to avoid contamination, fire or fume hazards, or loss of materials by evaporation. Closed tanks must be fitted with a vent pipe and those containing flammable liquids must be properly earthed and fitted with flame traps on all vents. Their contents may be measured by means of graduated sight glass, float gauge, level probe, or by the use of a dipstick (see Chapter 5).

In recent years increasing use is being made of rubber or plastic containers which may be collapsed to conserve space when not in use. This type has found particular application in the transport of large quantities of liquids by road.

Conventional containers for liquids are usually kegs, barrels, drums, tins, bottles, or carboys. Drums, which are generally cylindrical and of 40 or 45 gallons capacity, are supplied in materials similar to the large tanks and fitted with side-bungs, end-bungs, or both. Drums must not be pressurized or stored near sources of heat; they must be rolled using both hands and not kicked; placed in position with the bung uppermost; and, if on their side, chocks must be fitted to prevent them from rolling. Carboys are generally of 10 gal capacity. The common spherical, flat-bottomed type in glass is now often replaced by cylindrical or rectangular carboys constructed in plastic. Carboys must not be pressurized, rolled, or exposed to heat sources. It must be stressed again that unlabelled containers should NOT be used under any circumstances.

Conveyance of Liquids

The movement of liquids is much simpler than that of solids because they flow readily and can adjust their shape to fill any container, pipe, or vessel. They do present a particular complication, however, because many are potentially poisonous, corrosive, flammable, and explosive and their vapour may be equally dangerous.

Many liquids are stored in bulk in large tanks and for the convenience of storekeeping and distribution these are grouped together in a *tank farm*. Distribution is achieved by pumping along pipes to the respective process areas where it is metered. The correct procedure for drawing the quantities of each liquid (usually written as a notice adjacent to each meter) must be strictly adhered to.

Pumps. By far the most efficient means of conveying liquids is the use of pumps, and this is the method most widely used, particularly for continuous production. A wide variety of pumps is available and their design and application are discussed in Chapter 3. The choice of pump depends on many factors, which include the nature of the liquid to be pumped (viscous, flammable, etc.), the pumping rate required (gallons per minute), and the height to which the liquid must be pumped (the *head*).

The following general operational instructions must be observed:

Check that all valves over the whole pipe route, particularly at manifolds, are correctly set before starting any pump.

The outlet valve of a centrifugal pump should be *closed* before starting and opened as soon as the pump has reached full speed. On the other hand, the outlet valve of reciprocating and other positive displacement pumps must be *open* before starting.

No pump must be allowed to run dry and any overheating of motor, bearing housing, or casing must be reported immediately for safety reasons.

Guards must not be removed from any moving part of a pump.

Before starting any mobile pump, ensure that it is earthed and the flexible hoses are properly secured.

Gravity flow. This method is used wherever possible since it is safe; it requires no expensive equipment; and, by careful control of the outlet valve, the flow rate may be accurately controlled. Naturally, the basic requirement is that the outlet of the discharging vessel must be situated at a higher point than the receiving vessel (Fig. 2.6).

Siphons. A siphon (or syphon) consists of a flexible hose or tube bent in the shape of an inverted U where one arm is longer than the other. The shorter arm is placed below the surface of the liquid to be conveyed and the lower end discharges the liquid to another vessel. Flow will commence when the tube is completely flooded

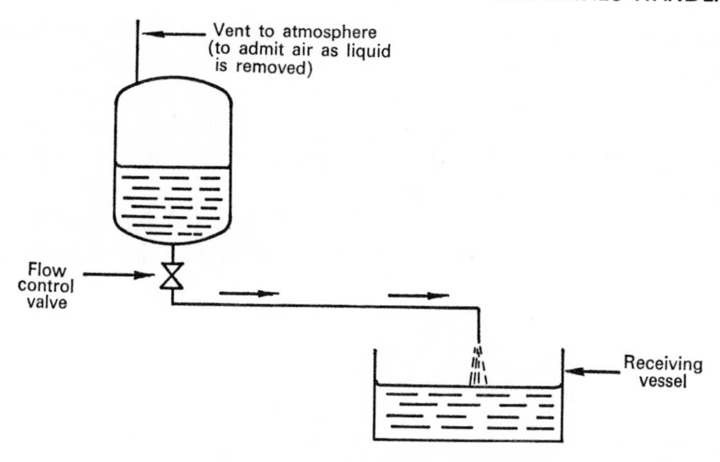

FIG. 2.6. Gravity flow of liquid.

with liquid and this is achieved by sucking at the lower, discharge end or by immersing the whole tube, sealing the ends and placing in the operating position (Fig. 2.7). Some siphons are fitted with a hand-operated suction pump at the lower end and a valve for

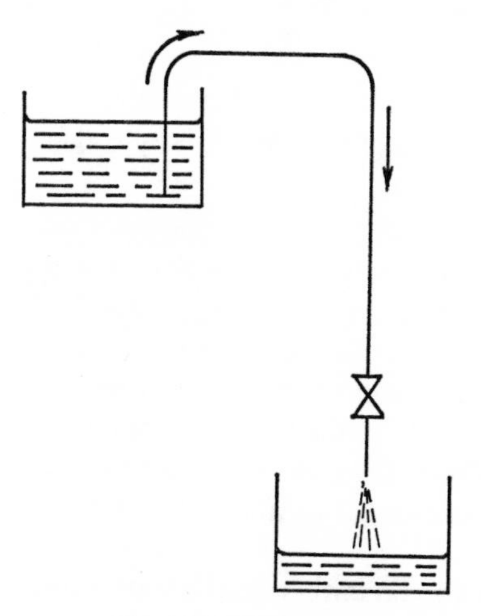

FIG. 2.7. Siphoning.

flow control. A siphon is only suitable for handling small quantities of non-corrosive liquids.

Pressure flow (*Airlift*). Basically, the method involves pressurizing a strong-walled vessel and discharging its contents through a submerged pipe or the bottom outlet.

The acid (or pressure) egg (Fig. 2.8) is used extensively in chemical factories, and, although not very efficient, it is particularly

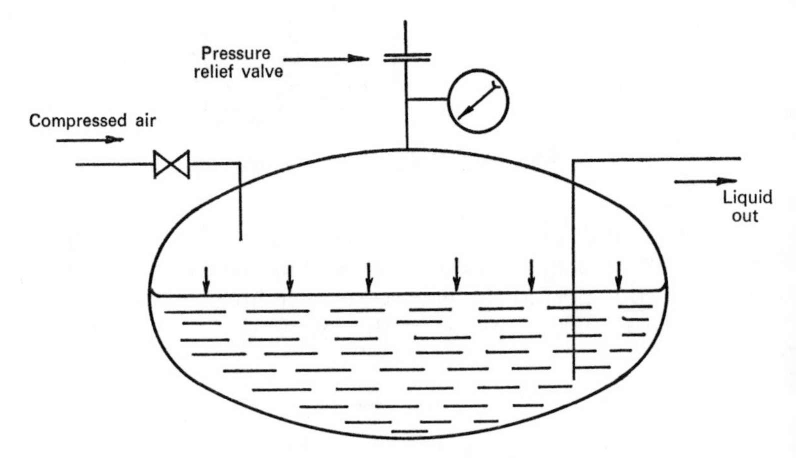

Fig. 2.8. Acid egg (pressure egg).

suitable for handling corrosive liquids or for intermittent conveyance of liquids. It consists of an egg-shaped vessel fitted with an inlet and an outlet pipe, one end of which is immersed in the liquid. Air is supplied to the egg through the inlet pipe and the pressure on the surface of the liquid forces it out of the discharge pipe. An operating valve is fitted to the inlet pipe and the vessel itself is fitted with a pressure gauge and safety pressure relief valve or a bursting disc.

It is important to remember that all pressure vessels are under mechanical strain and must be treated with care (see Chapter 8).

Vacuum flow. The technique is similar to pressure flow, except that in this case air is evacuated from the receiving vessel which then sucks in liquid via a dip-pipe in the discharging vessel. The sequence of operations, illustrated in Fig. 2.9, is as follows:

(i) Check that valves 1, 2, and 3 are properly closed.
(ii) *Slowly* open valve 1 (see p. 28, 139).
(iii) When vacuum gauge reaches a steady reading, *slowly* open valve 2 to admit liquid via suck-up line to receiving vessel at desired rate of flow.
(iv) Close valve 2.
(v) Close valve 1.
(vi) *Slowly* open valve 3 to re-admit air to vessel.
(vii) Open valve 2 to allow residual liquid in suck-up line to drain back to discharging vessel.

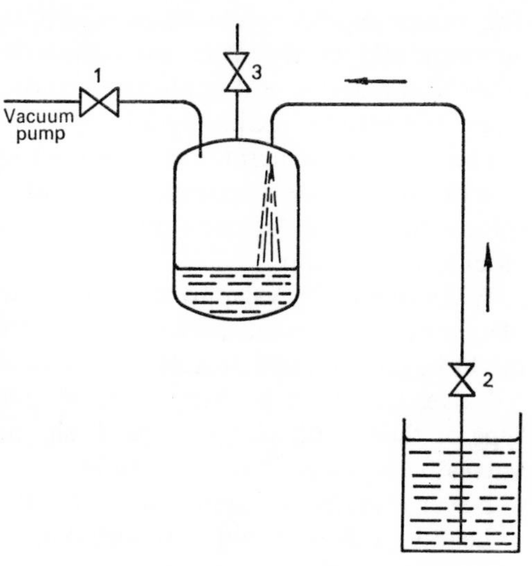

FIG. 2.9. Transfer of liquids by vacuum.

When handling a liquid which gives rise to a corrosive or flammable vapour which may damage the vacuum pump or constitute a hazard, *residual vacuum* may be used. In this procedure, when the vacuum gauge reaches a steady reading, valve 1 is closed before admitting liquid to the suck-up line.

Pipes and valves. No discussion of liquids handling would be complete without consideration of the selection of the necessary piping and valves. The satisfactory performance of any liquid handling depends on the material and design of the distributing pipeline and the type of valve which controls the rate of flow. The valves should be placed in positions easily accessible to the operator and to the fitter for maintenance. Reference may be made to the design and application of the various types of valves, pipes, and fittings in Chapter 3.

The following general operating instructions must be observed:

With very few exceptions, valves are opened by turning the spindle in an anticlockwise direction, and closed by turning in a clockwise direction.

When opening a valve to its fullest extent, make a final quarter of a turn in the opposite (clockwise) direction, otherwise the spindle might be jammed and mistaken for a closed valve.

Valves must not be overtightened, particularly the diaphragm type or those with glass-linings. It should never be necessary for valves to be more than hand tight.

Never attempt to "spin" valve spindles.

Leaking valves should be reported immediately.

HANDLING GASES

Storage of Gases

Bulk storage of gases at low pressure is normally in large cylindrical tanks called gasholders which may be seen at gas works. The gasholder may be constructed in sections which can telescope according to the quantity of gas being stored. Seals, to prevent the escape of gas at the base or at the section joints, are usually water or oil.

Smaller quantities of gas at high pressure are usually stored in bottle-shaped gas cylinders. Although convenient to handle, they suffer from the disadvantage that their thick steel walls make them very heavy. Cylinders must be held in position by securing chains and they must not be banged or exposed to sources of heat. Owing to the severe risk in the event of fire, only the minimum number of cylinders required for immediate use should be stored at the plant.

Conveyance of Gases

When discharging gas from cylinders, the protective cap is removed to gain access to the delivery valve. These valves, which are usually constructed in bronze or steel, are opened by turning *slowly* anticlockwise; except in the case of an oxygen cylinder which is opened in a clockwise direction. Many cylinders include a small needle valve which is specially designed to give a fine adjustment of a slow rate of gas flow. All cylinder valves should be kept free from grease. The cylinder cap must always be replaced after use and a label showing the cylinder to be empty firmly attached. Faulty cylinders must also be appropriately labelled to avoid risk to others and to ensure their replacement. Cylinders are painted in distinctive colours or given coloured bands around the neck for ease of identification.

Many cylinders contain liquefied gas which may be drawn off by connecting to a reinforced hose or line and inverting the cylinder. Strict safety precautions must be observed when carrying out this operation.

The transfer of large volumes of gas at low pressure is carried out by means of fans, whilst at higher pressures pumps are used. These pumps are similar to the liquid pumps but are modified to handle the larger volumes of gases and take into account their different physical properties (see Chapter 3).

Bulk storage of large quantities of gas in liquid form, for example oxygen, nitrogen, or argon, are becoming more common today. The gas is stored in an insulated pressure vessel and the gas is released through an evaporator and suitable control valve to a distribution main.

MATERIALS SAMPLING AND TESTING

Solids

Bulk solids usually exhibit considerable variation in quality throughout the material and special sampling methods must be used to obtain a small quantity for testing. The sample should be as representative as possible of *all* the material. It is a waste of time testing any material unless a sample is properly taken.

The simplest method, known as *grab sampling*, consists in taking many equal portions of the materials selected at random throughout the bulk. The more portions that are taken, the more accurate the sample; but even so this is not a very reliable method. The *coning or quartering* method is often used to reduce a grab sample to smaller proportions and improve its reliability. The material is piled into a flat-topped conical heap, split into four equal quarters, two opposite quarters are rejected, and the remaining material is gathered together in the shape of the original heap. The whole process is repeated until a sample of manageable size is obtained. This procedure may be extended further by the use of a *riffle*, consisting of series of shutes which distribute the material evenly into collecting boxes.

Another method, which is most suitable for finely divided powders, is *thief sampling*. The "thief" is a long, hollow tube which is thrust to the full depth of a container to fill it with material. After withdrawal the core sample may be conveniently discharged.

When sampling material which has been tray-dried, it is preferable to take the sample directly from the trays where the material is readily accessible, rather than after packaging. Portions should be selected from each corner and the centre of the tray, and the same procedure should be adopted for a rack of several trays.

Automatic samplers operating continuously at a predetermined time interval are often used, particularly for solids moving along belts.

When any sample has been taken, the container must be properly labelled to indicate the nature of the material, the quantity the sample represents, and its source, for example the batch number. It is then submitted to the quality control or analytical laboratory for testing. These tests will ensure that the material, whether raw material or product, meets the standard laid down in the quality specifications for its particular use. Common tests on solid materials include those for colour, water or solvent content, melting point (for organic chemicals which have relatively low melting points), dirt, appearance, bulk density, particle size, and assay (the percentage by weight of the active ingredient).

Liquids

Clear liquids are less difficult to sample than solids, since mixing is easier. However, mixtures of liquids which are immiscible (that is, they form two or more layers on standing) and those containing solids in suspension, present special sampling problems.

Where possible, liquids should be thoroughly stirred before the sample is taken. *Dip sampling* is carried out by lowering into the liquid a small container fastened to a long pole. The container is often fitted with a tight-fitting stopper which may be opened at the desired depth by means of a rod attached to it. Thief sampling may be used in a similar manner to that previously described; the lower end fitted with a valve.

Liquids flowing in a pipeline may be sampled continuously by fitting a pipe of small bore such that a portion of the liquid stream is diverted for the sample. The small bore pipe may be inserted through the wall of the main pipeline with special fittings at its end to obtain a more uniform sample over the cross-section of the liquid stream.

Tests commonly used to determine the quality of liquids are specific gravity, colour, clarity, refractive index, boiling point, freezing point, and viscosity.

Gases

Gases mix even more readily than liquids, in fact they mix spontaneously. Nevertheless sampling problems may arise by contamination with air or the presence of moisture or dust particles which may separate from the gas in the sample container.

A simple device for sampling gases is an aspirator, which usually consists of two bottles or bulbs connected through their bottom outlets by a flexible hose and containing a quantity of liquid which will not react with or absorb the gas. The right-hand bulb in Fig. 2.10 is lowered to admit the gas at valve A. Valves A and B are closed when the left-hand bulb is full of gas.

Gases flowing in a pipeline are sampled by methods similar to

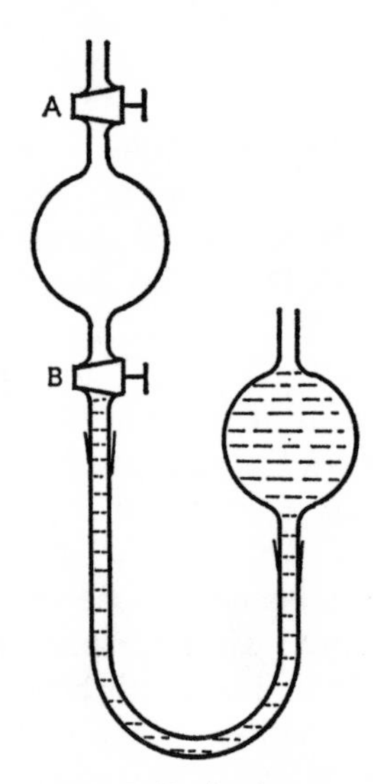

FIG. 2.10. Aspirator.

those described for liquids. Several types of probe are used to obtain a uniform sample over the whole area of the pipeline. Because of the likelihood of cavitation or pockets of stationary gas, it is unwise to take a gas sample near valves or other obstructions.

Gases are generally tested by forcing the sample through a series of bulbs or tubes containing liquids or solids which will absorb each component; for example alkaline pyrogallol will absorb oxygen, caustic soda or caustic potash will absorb carbon dioxide, anhydrous calcium chloride will absorb moisture, and so on. The quantity absorbed in each bulb is measured by the reduction in volume (at the same pressure) of the original sample. Flue gases are often sampled and tested automatically on a continuous basis.

CHAPTER 3

Chemical Plant

This chapter reviews the common items of equipment which, when suitably interconnected and provided with the necessary services, form a complete working unit of plant capable of performing one or more unit operations. It is recommended that the descriptions of equipment given in this review are read in conjunction with Chapter 4, so that their application in chemical processing may be better understood.

FLOW DIAGRAMS

To understand the design and function of a chemical plant it is a useful preliminary to study the flow diagram (sometimes called a flow sheet). Its purpose is to illustrate diagramatically, on one sheet of paper, all the items of equipment required for a chemical process or series of processes to be operated. Items are represented by simple symbols which are usually those recommended by the British Standard 974:1953 (your supervisor may have a copy). The route of materials through the equipment is indicated by arrows and interconnecting lines and, in some instances, the materials of construction, size or capacity, flow rates, power requirements, and pipe dimensions are also shown.

The flow diagram (Fig. 3.1) is built up from a description of the process which usually follows preliminary trials on a small-scale *pilot plant*. Initially it is used to discuss refinements and modifications to the process and its operation, engineering, and safety

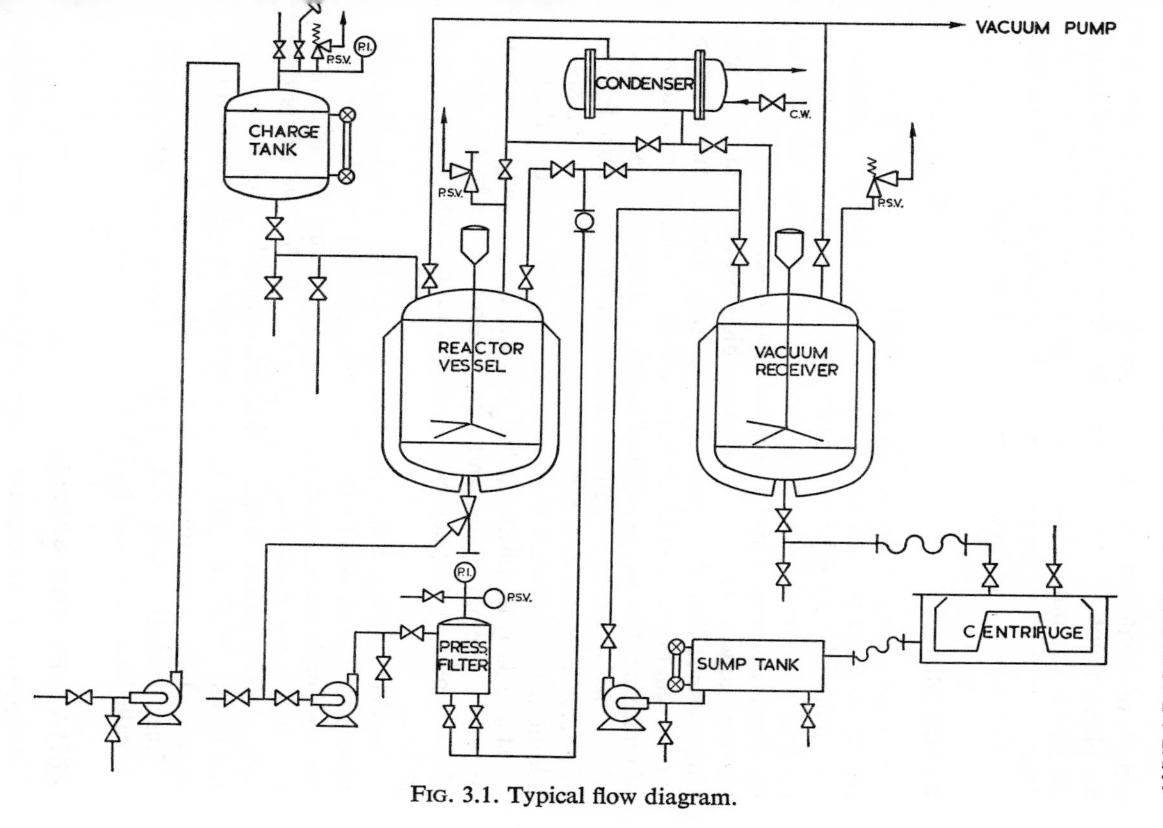

FIG. 3.1. Typical flow diagram.

aspects of the design. When it is finalized, it is used as a basis for the plant layout, plant numbering (flow sheet number), and drawing up an equipment list. When the plant is assembled the flow diagram is used to prepare operating instructions (batch sheets) and for the instruction of supervisors and chemical operators.

REACTORS

A reactor may be defined as any vessel in which the chemical reactions of the process are carried out. The required ingredients (raw materials) are fed into the reactor and may be mixed, heated, cooled, pressurized, distilled, etc., to bring about the desired chemical reactions. These vessels must be constructed to withstand the mechanical stresses of vacuum, high pressure, high or low temperature; they must have the necessary connections or additions to enable these conditions to be applied, and always a means of removing the final product of reaction. They may also be fitted with a mixing device or agitator, dip-pipes for sub-surface feeding of liquids or gases, and means of access for inspection and solids addition. The materials of construction must be such that they will resist attack by the chemicals fed to it and the products of reaction over the range of pressure and temperature dictated by the process (see Chapter 1). Capacities of reactors range from 10 to 50,000 gallons.

The details of the design of the reactor will be calculated for each process and these will dictate the capacity of the vessel, the size of the cooling/heating jacket or coil, the speed and type of agitator and baffle, the type of agitator gland, and the size of its drive motor; the position, range, and sensitivity of the instruments; the size of inlet and outlet nozzels, and the type of valves used in these nozzles. Typical reactors are shown in Fig. 3.2.

Reactors with strong walls, used for reactions at high pressure are called *autoclaves* (see p. 138.)

AGITATORS AND BAFFLES

Agitators or stirrers are essentially mixing devices used for handling liquids and liquid–solid or liquid–gas mixtures. There are five main types: propeller, turbine, impeller, paddle, and anchor, illustrated in Fig. 3.3.

The choice of agitator is dictated mainly by the type of mixing required, for example heat transfer, gas absorption, or emulsification, the suspension of solid particles in a liquid and the type of material being mixed, for example its viscosity, specific gravity, heat conductivity or size of the suspended particles. Generally, paddle, impeller, and anchor agitators are rotated at lower speeds (up to 150 r.p.m.) than the propeller or turbine type (up to 2900

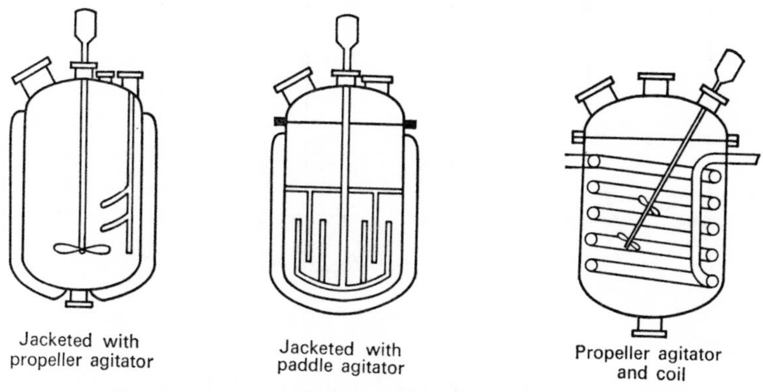

FIG. 3.2. Typical reactors.

r.p.m. in smaller vessels). The efficiency of agitation is considerably increased by the use of baffles, which break up the otherwise smooth, swirling motion set up by the agitator. The turbulence caused by breaking up the flow increase the mixing effect of the agitator, reduces cavitation and the tendency for the whole vessel to vibrate or rock.

Standard baffles fitted to glass-lined vessels are the finger type (shaped like an inverted F), the d-type, and, more recently, the h-type.

TANKS

Most chemical processes utilize tanks of various shapes and sizes, which are named according to their particular function.

They may be called head tanks, charge tanks, sump tanks, holding tanks, etc., but basically they have the common purpose of providing storage capacity. They are usually plain cylindrical or rectangular vessels, fitted with various inlet nozzles, a draw-off nozzle, probably a separate drain, a gauge (or sight) glass, and sample nozzles. The construction of the tank is determined by the material or materials being handled, the pressure or vacuum required, etc., and its size by the quantity required to be stored (see Chapter 1).

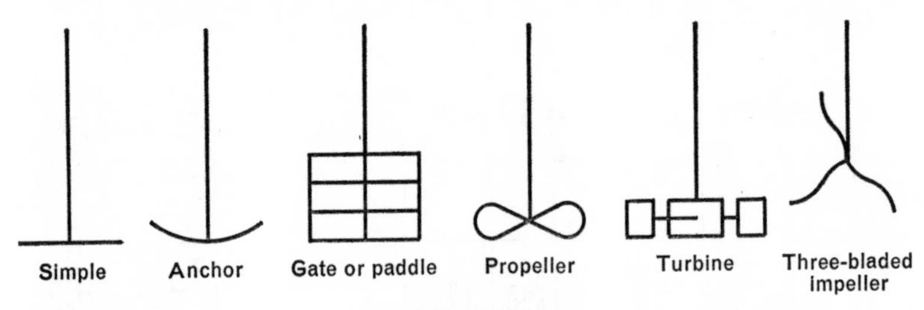

FIG. 3.3. Agitators.

TOWERS

Towers, or columns, are used in a variety of chemical processes, principally distillation, scrubbing, washing, separation, and gas or liquid absorption. It is usually required that some form of *packing* be used to increase the surface area of a liquid presented to a gas, when moving in opposite directions (see Chapter 4). The tower must, therefore, be constructed so that the packing is supported without being crushed under its own weight and the supports leave a large enough area for free passage without permitting the packing to fall through. Some towers are built with several supporting plates and with a layer of coarser packing immediately above each plate.

Inspection ports are usually fixed to the side of the tower to permit inspection of the condition of the packing and to allow it to be "topped up", because packing often settles down in a tower after a short initial period of operation. In some cases the materials of construction make conventional packing support difficult; for example, in a lead-lined tower (or scrubber) the coke packing is supported on acid-resistant brickwork.

CALCINERS AND KILNS

Both calciners and kilns used in chemical industry are usually rotary tubes slightly inclined and constructed of heat-resistant stainless steel or other special metal, or refractory brick-lined steel. The material is fed in at the higher end by screw conveyor or gravity chute, and a source of heat, for example producer gas or fuel oil flame, at the lower discharge end. In some cases the flame will be ignited in a separate combustion chamber and the hot gases only fed to the kiln. In other cases it may be necessary to keep even the hot gases out of the kiln, and indirect heating is used by passing the rotary kiln through a stationary furnace, the heat being transmitted through the walls of the inner rotating chamber. The length of the latter type is limited to the distance between the supporting rollers, but in directly heated kilns no such restriction exist and lengths of 180 ft are common.

CRYSTALLIZERS

The simplest form of crystallizer is an open steam-jacketed pan into which the saturated solution is poured and the solvent (usually water) is evaporated. To eliminate overheating of the product in contact with the heated wall a slow-speed, scraper-type rake rotates inside the pan. This also ensures uniform cooling and crystal size.

Larger continuous crystallizers often achieve crystal formation by cooling a supersaturated solution in a reactor-type vessel with brine or cooling water circulated through its jacket. The crystal slurry is pumped to a filter and the filtrate returned to a dissolver. In some cases crystallization is helped by cooling under conditions of vacuum. There are numerous special crystallizers designed to meet particular needs: for example the Votator rotating shaft crystallizer and the Passburg rocking crystallizer.

C.P.O.—D

HEAT EXCHANGERS

Any equipment in which heat is transferred between liquids, or liquids and gases through a wall which separates them may be described as a heat exchanger. Hence all jacketed vessels come within the wider definition of heat exchangers, but in this section only the items that are designed specifically for "exchange of heat" will be mentioned. These may be called recuperators, condensers, heating bundles, etc., according to the particular application of the heat exchanger. The detailed design of a heat

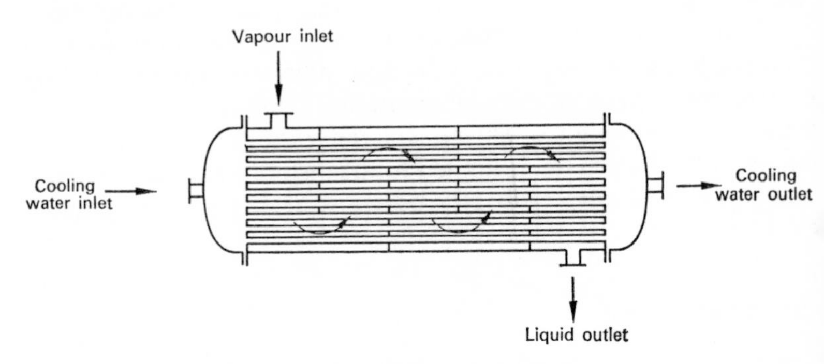

FIG. 3.4. Single-pass condenser.

exchanger is influenced mainly by the type of liquids being used, the temperature difference between them and the quantity of liquids flowing.

With the single-pass condenser, illustrated in Fig. 3.4, the coolant (usually water) passes through the tubes and the hot vapour is fed into the top of the outer shell. Condensate drains away from the bottom of the shell.

It is possible to arrange the tubes and the "header" or end covers so that the water enters only one bank of tubes from the "inlet header" and returns down another bank from the opposite header. Each passage through the tubes is called a *pass*. Four-, six-, eight-, and twelve-pass condensers are quite common (Fig. 3.5).

In addition to the numerous arrangements of coolant passes there are often baffles inserted inside the shell which deflect the vapour to increase its velocity and ensure full use of the tube area available. In some instances the condensate outlet and the coolant outlet are adjacent, called *cocurrent;* in other types the condensate outlet is adjacent to the coolant inlet, called *countercurrent*, giving the coolest condensate.

The velocity of the fluids is often important as it aids heat transference by forced convection. The efficiency of a condenser is influenced by the cleanliness of the interconnecting surfaces and

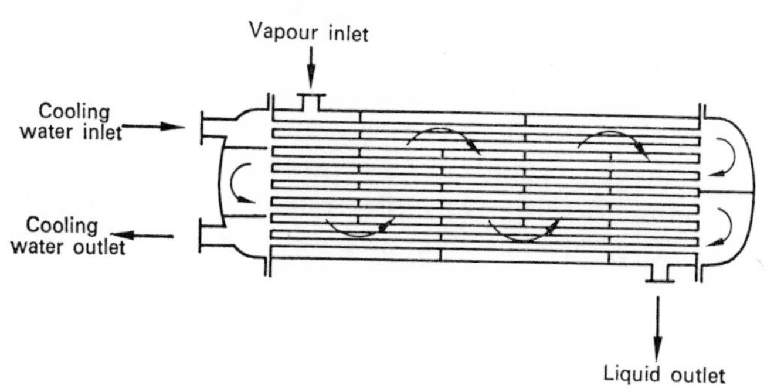

FIG. 3.5. Four-pass condenser.

it is important that the coolant does not leave a deposit (see Chapter 7, under water). On the vapour side this is also important, but it often is more important to construct the condenser of a resistant material. Graphited carbon is used very successfully on this duty as it is a very good heat conductor and also very resistant to most chemicals. Often the familiar shell and tubes are replaced by a carbon block in which holes are drilled to allow vapour to pass one way and the coolant to pass through similar holes drilled at 90 degrees to the former: the resultant heat exchanger is very compact. Special plate-type heat exchangers which achieve a high degree of turbulence and high heat-transfer coefficients are available: they are mainly used where small temperature differences are required.

Recuperators are usually employed to heat an incoming fluid with waste heat from a process. For example a waste-heat boiler is, in principle, a heat exchanger using the waste heat from probably a blast furnace to evaporate water to steam. In other cases recuperators heat incoming fluid and cool outgoing fluid simultaneously. On a steam-generating plant flue gases are often used in economizers to heat the feed water to the boiler.

Heating bundles are heat exchangers fitted to the bottom section of a still or reboiler, through which steam or hot gases are passed to increase the temperature of the feed. These are often "hairpin" tubes making one loop so that the bundle requires only one connection in the still.

Another form of condenser frequently used in chemical industry is the *spray condenser*. This is a modified form of spray absorber or scrubber in which cold water is sprayed into a stream of vapour to cool it and to flush away the resultant condensate. This is usually employed in connection with water-sealed vacuum pumps.

DRIERS

The most important types of batch drier which will be considered are:

(a) Atmospheric
(b) Vacuum
(c) Rotary
(d) Fluid bed.

(a) *Atmospheric driers* are usually of the tray and rack type; the product being spread on trays which are loaded onto racks and positioned in a cabinet around which hot air is circulated. The heat is usually obtained from a steam coil and the hot air is circulated by one or more fans. The cabinet incorporates an air inlet and outlet fitted with damper controls and louvres in front of the fans. The latter ensures an even distribution of the circulating air and the damper controls enable the humidity of the air to be maintained at an optimum level. Clearly, in the early stages of drying, a high proportion of fresh air should be introduced, but as the drying continues heat can be conserved by closing the damper and recirculating more and more of the hot air. The temperature of the air is usually thermostatically controlled. These driers are very reliable on any duty that does not produce an acidic atmosphere, and for chemicals which are not heat sensitive.

(b) *Vacuum driers* make use of the principle that moisture will vaporize at a lower temperature under vacuum than at atmospheric pressure. Therefore they are used where excessive heating may scorch, discolour, melt, or chemically decompose the product.

Simple vacuum driers are constructed of heated plates (on which trays containing wet material are placed) fitted in a heavily built metal cabinet with a tightly fitting door. Vacuum is applied by an external pump which often requires a condenser and sump tank to protect it. Generally, the heat transfer to the trays is poor because of uneven contact surfaces and, since under vacuum conditions most of the heat is obtained by conduction, this has a marked effect on the efficiency.

Modern designs of tray vacuum driers usually incorporate the heating plate in the tray; large trays with a jacket along the underside are heated by vacuum steam and placed in individual heated compartments to which vacuum is applied. This design is advantageous when small quantities of different materials are to be dried simultaneously as each compartment can be isolated from the others for loading and unloading.

(c) *Rotary driers*, in their simplest form, are horizontal cylinders provided with a door for charging and discharging material. They are also constructed in spherical or double-coned shapes. The unit may be jacketed or heated by an internal cylinder with the outer casing itself rotating, and operation may be either at atmospheric pressure or with vacuum. For continuous operation the drier may be heated by a direct flame on the rotating shell, with the wet material being fed to the lower end, carried up by blades along the inner walls, and discharged at the upper end.

In a rotary louvre drier, hot air is blown through a series of louvres in a double-walled rotating cylinder and up through the wet solids.

Rotary driers are available in a variety of metals and in glass-lined steel to handle corrosive materials. They are particularly useful for drying material of poor heat conductivity and uniformly blending the material during the drying cycle. They suffer from the disadvantage that materials which are very wet or will not break down easily into powders may form a hard cake on the walls or form into balls.

(d) The use of *fluid bed driers* is restricted to material that will fluidize. Although designed primarily for water-wet materials, they may be modified to handle organic solvents. The wet material is spread over a bed of fine woven metal or nylon mesh and heated air or gas is blown through the material at a fairly high velocity. The fluidization thus brought about gives intimate contact between the air and the suspended particles of the material. The drying cycle is considerably reduced, the flow area required is smaller, and material handling is reduced.

EVAPORATORS

Evaporators are steam-heated, jacketed pans which supply the heat required to evaporate the solvent from the solids in suspension, or from another liquid with a lower boiling point. When evaporation under vacuum is required the pan is fitted with a cover and vacuum is applied. These pans are available in most materials of construction and are used mainly for evaporating the solvent from the suspended solids. The remaining solids are then dug out of the pan.

Calandrias are evaporators in which the liquid is exposed to a large heated area by building a bank of tubes into the lower half of a vessel and passing steam outside the tubes. The solution is heated and circulated up the tubes, the vapour separating from the lower boiling point liquid in the upper half of the vessel. Baffles are built into the vapour take-off and the liquid droplets collect and drain back to the base of the calandria. The major problem with this type of evaporator is the blocking of the tubes and for this reason an external heat exchanger is often used; the circulation is maintained by using a suitable pump.

The water-cooling tower described in Chapter 7 is a very common example of an evaporator.

FILTERS

The older filters consisted of a timber frame supporting a cloth placed above a suitable storage tank. This method was extremely slow and nowadays the application of pressure across the filter cloth is usually employed to speed up the operation. *Plate and frame filters* (Fig. 3.6) which employ this principle are usually constructed of timber, cast iron, or rubber-covered steel. The plates and frames have the same outside shape, the plates support the filter cloth against a ribbed surface and the frames accumulate the solids. The number of plates and frames used may be varied, and their order of assembly is often indicated by markings stamped on the outer edges. The slurry is fed into the press by a pump and when the specified pressure is reached the press is full. With this type of filter the cake can be subjected to compressed air which will remove most of the filtrate and dry the cake to a friable state. This enables the frames to be emptied readily when pulled apart. There are various arrangements of feed holes in the plates which allow for sealed collection of the filtrate, cake washing and cake blowing.

Enclosed *pressure filters*, usually circular with horizontal or vertical plates, have the advantage that they are more compact and give a better distribution of filtered solids. The plates in these filters are usually fabricated and assembled as a cartridge which is compressed before being lowered into the filter case. They are available in many materials of construction, for example stainless steel and rubber or plastic-lined metal.

Leaf filters, like pressure filters, are enclosed in a vessel and are compact. The filter leaf is a frame over which a filter cloth is stretched, the slurry being fed to the outside of the frame. Similar

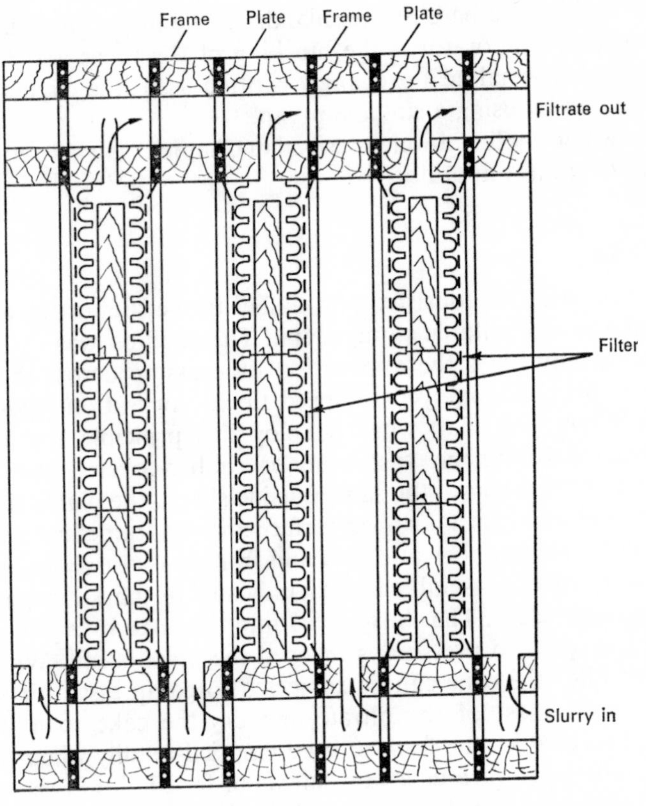

FIG. 3.6. Plate and frame filter.

to pressure-plate filters, they can be fabricated in most materials of construction.

Centrifuges, sometimes called hydro-extractors or "spinners", are another very common form of filter. They consist of a basket or cage, mounted horizontally or vertically, with holes in the side, which is rotated at a speed such that the centrifugal force throws the contents outwards. In a domestic "spin drier" the holes are large enough to allow the water to pass but not the clothes. The same principle is applied with a slurry but a filter cloth is fitted inside the basket to retain the solids in the suspension. It is usual to fit a coarse-wire gauze behind the cloth to increase the area available for the liquid to be thrown out.

Centrifuge baskets are carefully balanced by the manufacturer and it is important that the feed is distributed evenly or the machine will vibrate violently. Often they are fitted with some form of variable speed drive so that the feed can be achieved at a suitable rate to ensure even build-up of cake and the final spin at a higher speed will remove as much water as possible.

The spinning basket has a large momentum and acts like a flywheel, so that when the motor is switched off it will continue to run for a long period. Safety interlocks are fitted so that the lid cannot be opened whilst the machine is rotating and that the application of the brake will cut out the motor. Similarly, there is a cut-out switch which will operate when excessive vibration occurs.

Centrifuges require flexible connections for feed, wash, and discharge lines and, because it is difficult to seal, its use where toxic vapours arise should be avoided. The discharge of a centrifuge is done either by hand, with scoops or small shovels, or mechanically, by a plough inside the basket which scrapes the cake away and discharges it through a hole in the bottom of the basket.

There are many types of small filter unit which are fitted in pipelines for removal of fine particles or traces of solid material which may have by-passed a larger filter unit. They are often referred to as *polishing filters*, and sometimes called *cartridge* or *line filters*. They consist of an enlarged tube, placed at 90 degrees to the pipeline, housing a filter medium. This medium is usually in the form of a hollow cylinder, called a cartridge or element, which may be of spun cotton, glass fibre, sintered bronze or steel, pleated, resin-impregnated cellulose, or synthetic fibre, such as polypropylene. The liquor enters the outer casing and passes through the cartridge wall to the inner hole and thence to the outlet (Fig. 3.7). An alternative design incorporates metal discs, accurately spaced to retain solid particles as small as a few microns in diameter.

MIXERS

Mixing of liquids has been mentioned earlier in this chapter (p. 36) when agitation was discussed. Mixing of powders or pastes is achieved in trough-like vessels fitted with slow-moving spiral blades or rakes. Usually, these are batch mixers, which are loaded from the top and discharged through a suitable control valve at the bottom. The time required to achieve homogeneity

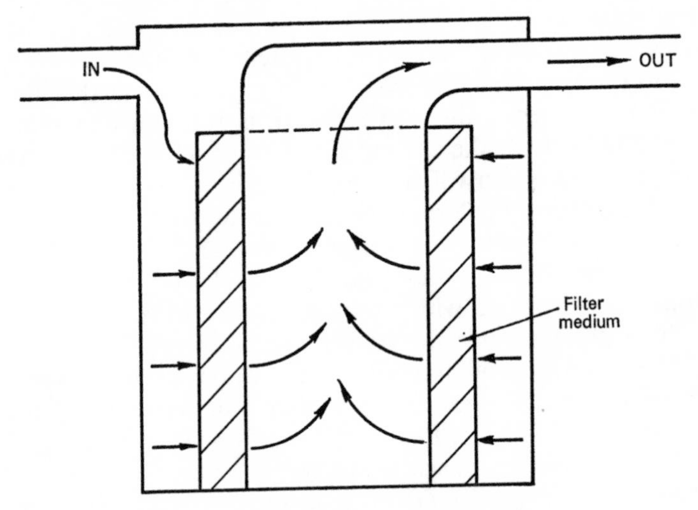

FIG. 3.7. Polishing filter (cartridge filter).

will depend on the free-flowing characteristic of the powders and additives. Continuous mixing can be achieved in some screw conveyors which are fed at one end and discharged at a higher level at the other (see p. 19).

MILLS AND CRUSHERS

There is a vast range of equipment designed to achieve size reduction of material. Large boulders are usually broken down in *jaw crushers*, which will accept lumps up to 12 in. in size at rates of approximately 10 ton/hr and discharge the crushed material at under 1 in. size. Crushing is achieved by compressing the lumps between the fixed and moving jaws, the latter being pivoted at the discharge end, and driven by an eccentric shaft. This ensures that there is a maximum size that the crusher will discharge. Some machines incorporate a rolling action which aids discharge and reduces clogging. The eccentric action puts a fluctuating load on the drive and to cushion this the drive will incorporate a large flywheel. These crushers are noisy, cause considerable vibration, and often give rise to large quantities of dust.

Cone crushers are constructed so that a cone-shaped jaw, large at the discharge end, is driven eccentrically inside a conical case that is larger at the feed end. This gives a continuous crushing action and reduces the vibration; it also has a higher throughput.

For the size reduction of smaller particles, *hammer mills* are often employed. In these machines, free-swinging hammers are rotated at high speed so that they are thrown out by the centrifugal force, hit the feed, and break it down by the force of impact. In many cases the discharge from this mill is through a sieve or screen to control the output particle size.

Another form of crusher incorporates twin rollers rotating in opposite directions so that the feed is drawn between them by friction and the weight of the feed. One of the rollers is spring-loaded to eliminate excessive overloading of the drive. Similar crushing can be achieved in *pan mills*, consisting of a pan above which two heavy rollers are mounted. Usually these rollers, which are freely pivoted and driven around the pan, crush the product by their weight. In some types the pan rotates and the roller pivots are stationary, whilst in others both the pan and the rollers rotate in opposite directions. In all cases ploughs can be fitted to direct the material into the path of the rollers and also into the discharge chute. Water and/or lubricants can easily be sprayed onto the product.

Fine milling is achieved in a *ball or rod mill*. It consists of a large cylinder rotating about its longitudinal axis, containing a number of loose, hardened steel balls or rods. These are carried up the side of the rotating cylinder and milling is achieved when they fall over each other in the material. The length of the mill and the shape determine the rate of throughput and the product size.

Fine powders are produced in *pulverizers* or *micropulverizers*, which are basically hammer mills running at a high speed and incorporating a centrifugal fan which enables an air sweep to aid the classification and reduce the particle size to approximately 0·005 mm. Pulverizers require a high-powered drive and facilities for discharge into dust collectors and bag filters.

PUMPS

There are three main classifications for pumps used in the chemical industry: centrifugal, reciprocating, and rotary.

Centrifugal pumps are the most widely used for liquid pumping because they are capable of transferring large volumes without any dependence on valves or fine clearances and can be run against a closed valve without developing a very high pressure. The design can be simple so that lining of the volute and covering of the impeller with any required material is possible; for example rubber or plastic or glass-lined pumps are readily available. Furthermore, centrifugal pumps are most versatile in that they can handle a wide range of slurries or solids in suspension, in addition to liquids with high viscosities. The principle of operation is to draw fluid into the centre of the casing and sweep it towards the periphery by blades spinning at high speed.

A very simple centrifugal pump is shown diagrammatically in Fig. 3.8. This would be the basis of designs to be used in lined pumps or those handling heavy slurries or suspensions. Generally they would not be very efficient, but can be improved by shrouding the blades and reducing the clearance between the impeller and the casing.

The main disadvantages of centrifugal pumps are the limitations of delivery pressure and their inability to prime themselves. The former can be overcome by using twin or multi-stages, usually on the same axis, and driven by one motor. The fitting of a self-primer, described in Chapter 6, will eliminate the latter disadvantage.

Reciprocating pumps of all types displace the liquid or gas by the movement of a piston inside a cylinder and the operation of non-return valves on the inlet and outlet of the pump.

A reciprocating (plunger) pump, illustrated diagramatically in Fig. 3.8, will operate for very long periods with the minimum of wear and be trouble free; it is necessary, however, to maintain a close fit between the plunger and the cylinder and to ensure that the ball valves seal firmly. The volumetric efficiency of these pumps is high but machined finishes required in the cylinder, on the piston, and the valve seats are likely to deteriorate rapidly when handling

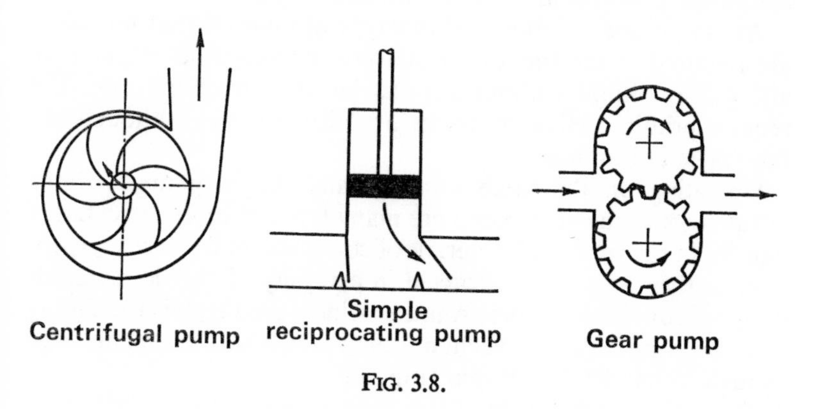

FIG. 3.8.

corrosive liquors. Furthermore, any suspended particles or dirt in the liquid would be likely to prevent tight sealing of the valves and cause wear by abrasion.

A reciprocating pump, when handling non-corrosive fluids, is capable of developing high pressures or producing vacuum. Like the centrifugal pump, it can be arranged multi-stage so that one cylinder discharge is connected to the suction of another. This type is often employed as a vacuum pump or as an air or gas pump (usually called a compressor), and in many cases as a metering pump because of the fixed volume discharged per stroke.

There are many types of valve used in reciprocating pumps: simple dead-weight balls, slide valves, spring plates, or poppet valves. It must be emphasized that reciprocating pumps are designed for particular duties and are not versatile. A pump or compressor designed to handle a gas would be arranged with a very small clearance above its piston and would not be capable of handling relatively incompressible liquids.

Rotary pumps are dependent on some form of slide valve capturing a quantity of fluids and forcing it through the pump casing. The most straightforward type is the *gear pump*, the meshing of the gears forming the slide valve and the fluid being carried around the pump casing between the gear teeth and being discharged before the gears mesh in the centre (see Fig. 3.8).

An important advantage of this type of pump is that no valves are required in the suction or delivery, it is capable of pumping air, gas, or liquid without any detrimental effect and does not require priming. High pressures are also possible, although the flow rates are limited.

Operating on the same principle and retaining most of the advantages of the gear pump are many types of eccentric cylinder pumps. In conventional materials of construction they are inexpensive, or they may be constructed in a variety of plastic materials designed for special duties. Another widely used type is the sliding vane air compressor or vacuum pump which is often made large enough to handle 700 ft^3/min (Fig. 3.9).

The main disadvantage of this type of pump is the close tolerances required between the ends of the rotors and the casing which are susceptible to deterioration when corrosive liquors or vapours are being handled.

VALVES

The main classifications of valves are gate, globe, ball, and plug, diaphragm, damper, and butterfly.

The *gate valve* (Fig. 3.10) is probably the most straightforward type; the pipeline is closed by pushing a gate across the pipe, actuated by a screw thread and guided by a U-shaped seat. It is primarily an on or off valve and is not ideally suited to regulate the flow in the pipeline. Difficulty in obtaining a tight closure will be experi-

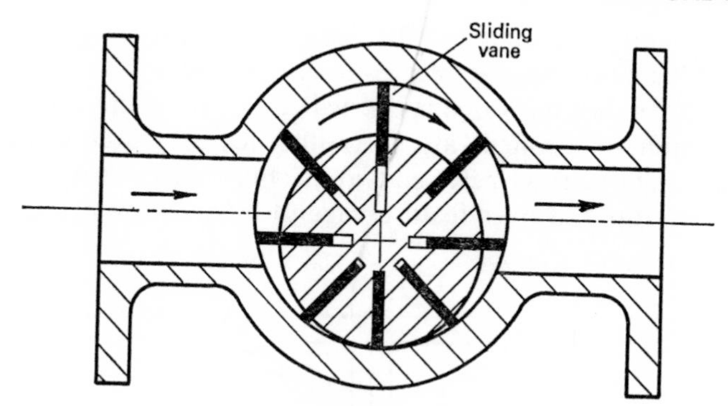

FIG. 3.9. Vacuum pump.

enced if small particles are trapped between the gate and the seat. Therefore it is inadvisable to use it in a slurry line, although some gate valves are specially designed to operate in these conditions, for example those in which the gate rolls into the closed position sweeping the seat clean at the same time.

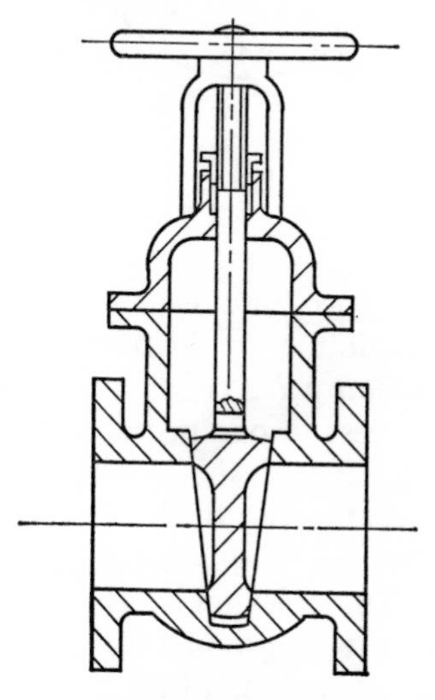

FIG. 3.10. Gate valve.

Globe valves are designed for controlling flow and are often fitted with shaped plugs or sleeves to give a flow rate proportional to the extent that the valve is opened; they are ideally suited for use as remote-controlled valves. Another feature, often incorporated, is a tight shut-off between two machined faces, but it always requires an up and over feature which obstructs the draining of

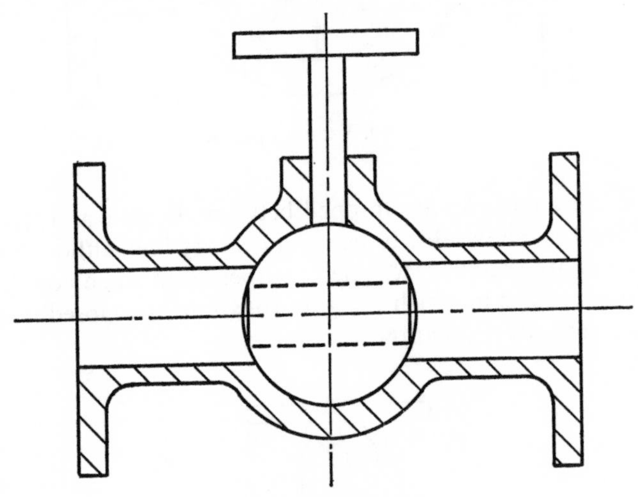

FIG. 3.11. Ball valve.

horizontal lines. The reliance for closure on machined faces limits their use with corrosive liquors or slurries.

The operation of a *ball valve* (Fig. 3.11) is the rotation of a sphere through 90 degrees; in one position the hole through the axis of the sphere is in line with the pipe (the open position) and in a position at 90 degrees to this the hole is across the line of the pipe (the closed position). It is quick acting and cannot be used for fine flow controls. The hole can be made as large as the pipe bore, and is then self-draining and roddable. It is available in many materials of construction, often including a PTFE ball. The plug cock valve (Fig. 3.12) is very similar to the ball valve, the

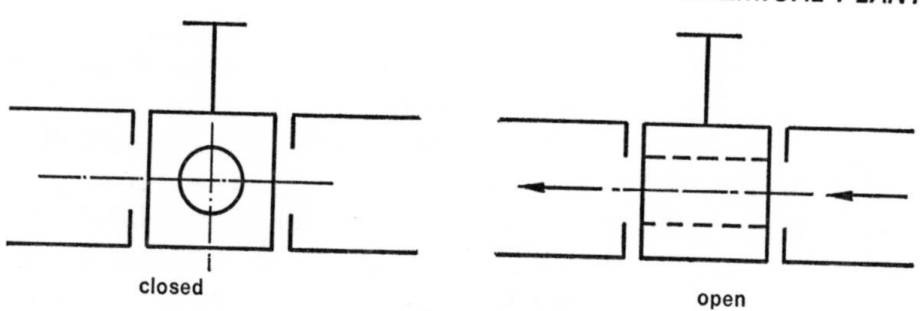

FIG. 3.12. Plug cock valve.

sphere being replaced by a cylinder or truncated cone and the hole is usually slotted.

Both types can be made in multi-port form, that is the hole in the ball or plug may be T or L shaped and the valve body has three or four outlets (Fig. 3.13). The careful selection of a single, multi-port valve will not only eliminate two or three valves but it will also eliminate the dead leg of material that remains in a branch above the conventional valve (Fig. 3.14).

Diaphragm valves are described more fully in Chapter 6. It will suffice to mention that the diaphragm eliminates gland leakage; they are available in almost all known materials of construction. Similarly there are numerous diaphragm materials available to

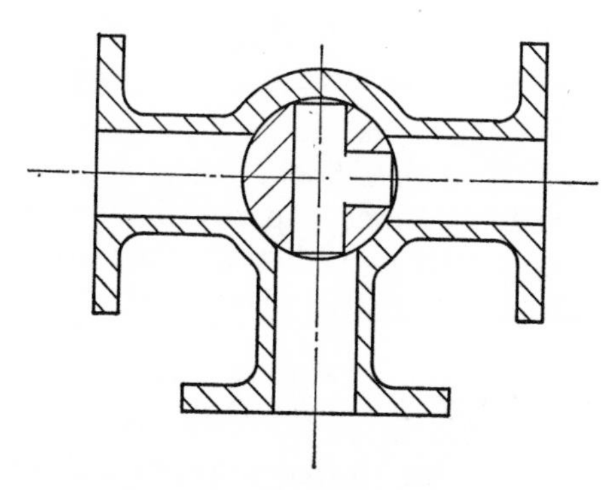

FIG. 3.13. T-Port ball valve.

match most requirements, including PTFE, which is often used as a standard. For these reasons it is not surprising that diaphragm valves are so widely used in chemical industry.

Glass valves, fitted with PTFE, glass impregnated plugs attached to PTFE bellows, are made with the same end-fittings as glass pipes. The plug and bellows are actuated by a handwheel and these two units can be removed without taking the body out on the pipeline. They have the advantage that the body is equally resistant

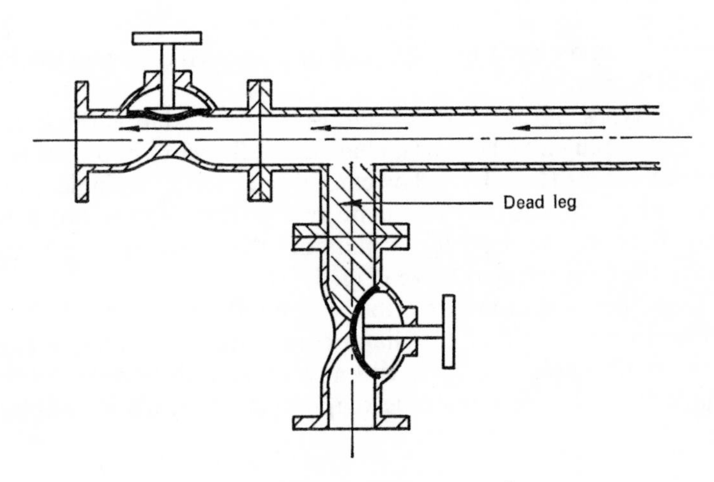

FIG. 3.14. Tee branch and dead leg.

to chemicals as the glass pipeline itself and can be inspected for damage through its own glass body. The plug and bellows have a very wide chemical resistance and overtightening is eliminated by a built-in torque limiter in the handle. Furthermore they do not require as elaborate supports as glass-lined valves do in similar positions.

Pinch valves are essentially a short length of rubber or plastic hose which is supported in such a way that it can be flattened between two bars that are moved towards each other when the operating hand wheel is turned or compressed air applied. It is a

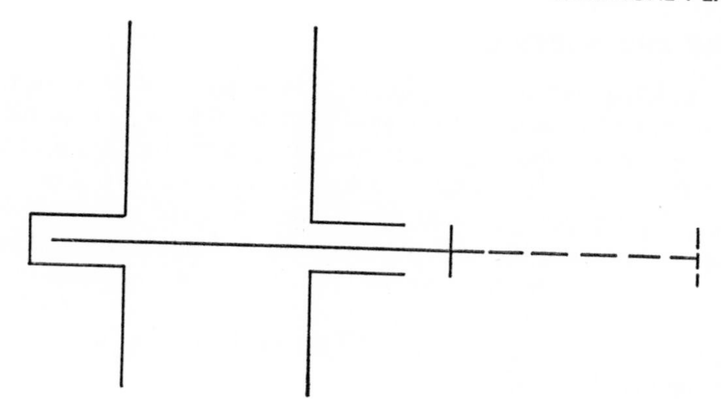

Fig. 3.15. Damper or slide valve.

very simple and trouble-free valve but it is limited to those chemicals that do not attack the tube material.

Dampers and butterfly valves (Figs. 3.15 and 3.16) are usually used in ducting or on the discharge from hoppers, blenders, etc. Their operation is straightforward and they are often a very loose fit in the pipe or duct. There are, however, more expensive types that are built to close tolerances and are capable of maintaining a good control of throughput. They are available in most materials of construction, including wood, metals, plastics, and rubber.

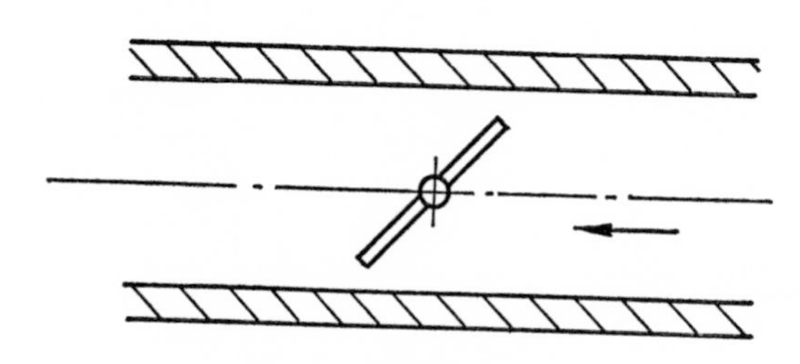

Fig. 3.16. Butterfly valve.

FANS AND DUCTING

Fans are designed for moving large volumes of air or gases at a relatively small pressure drop, usually up to 10 in. w.g. Their main use is in air-conditioning and ventilating systems, but they are also used for producing draught for burners and the conveyance of gases. The "pipes" through which fans blow are usually called ducts and because they are required only to contain a gas at low pressure, a light metal or plastic construction is frequently used. In some instances propeller fans are mounted in a wall, roof, or window to extract air or fumes. Further discussion of fans and ducting may be found in Chapter 6.

CHAPTER 4

Unit Operations

CHEMICAL operations may be classified as techniques of operation, specialized operations, and unit operations.

Techniques of operation relate to the principles and practice of materials handling; sampling and testing; measuring, for example weight, length, or volume; the use of instruments, for example in remote control; the use of services, such as steam, cooling water, or compressed air. Each technique is applicable to a greater or lesser degree to all chemical operations involving solids, liquids, and gases. Discussion of them may be found in Chapters 2, 5, and 7.

Specialized operations, such as catalysis, electrolysis, sterile and aseptic operations, and fermentation processes, are beyond the scope of this book and the reader should refer to special books on these subjects.

Basic unit operations may be defined as those by which chemicals undergo physical change, such as a reduction in particle size, a change in crystal form, or the change of a liquid to vapour. Unit operations may be carried out on a batch basis (i.e. discontinuous production) or on a continuous basis. It is important to distinguish between unit operations as described, and unit processes; such as esterification, nitration, or sulphonation, which are chemical changes by which new compounds are formed. They are, however, interrelated since no unit process can be carried out in the chemical factory without the application of one or more unit operations.

The basic unit operations are listed in Table 2, showing the type of chemical plant used.

TABLE 2

Unit operation	Chemical plant
Absorption and scrubbing	Absorbers, scrubbers, towers
Calcination	Calciners, kilns
Cooling and refrigeration	Coolers, refrigerators
Distillation and fractionation	Stills
Drying	Driers
Evaporation	Evaporators
Filtration	Filters
Mixing	Mixers
Precipitation	Precipitators, general chemical plant
Size reduction and size separation	Mills, crushers, screens, classifiers
Solvent extraction	Extractors, mixers, settlers

The nature of some of these operations is now discussed. The reader should also refer to Chapter 3 for details of the chemical plant used.

ABSORPTION AND SCRUBBING

The absorption of gases may be defined as a unit operation in which one or more components of a mixture of gases are removed by contact with a liquid. It is widely used in the chemical industry to extract a valuable gas from a mixture of gases or for the removal of noxious components such as hydrogen sulphide or sulphur dioxide from flue gases. The latter application is generally called *scrubbing*. Absorption is sometimes called gas–liquid extraction and is similar to solvent extraction (see p. 75). In each case, design of plant is concerned with the intimate mixing of the components, usually on a continuous basis, to achieve efficient extraction.

Absorption normally depends on the gas it is required to separate being dissolved in the liquid, but in some cases a chemical reaction may take place between the gas and the liquid. The choice of solvent depends on the quantity of gas it will dissolve, its cheapness, and the absence of hazardous properties. Water is used whenever possible and is particularly desirable for those gases which readily dissolve in it, for example hydrogen chloride or ammonia. The concentration of dissolved gas may be increased by

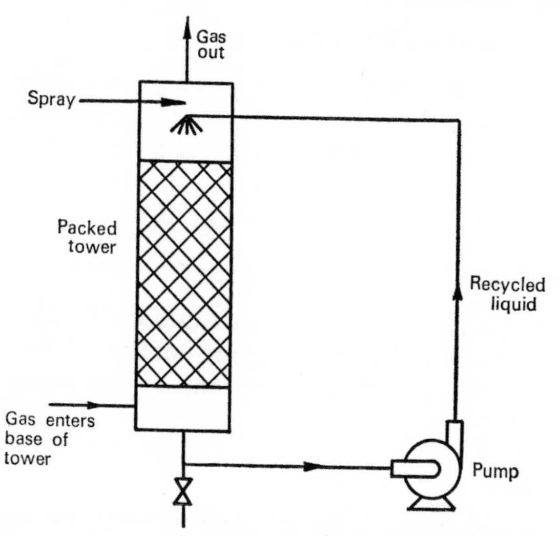

FIG. 4.1. Flow diagram for simple gas absorption using packed tower.

operating above atmospheric pressure; this particular method is used in the refining of petroleum.

An absorption process may be followed by *stripping*, in which the dissolved gases are regenerated in gaseous form for further use.

Packed Towers or Columns

This is the most common method of gas absorption (Fig. 4.1). The tower is similar to that used in distillation or solvent extractions and may be packed with Raschig rings, Berl saddles, or other types of packing (see Distillation, p. 65). The liquid is sprayed into the top of the tower and, on descending, meets a counter-current of the gases passing up the tower. The liquid containing the dissolved gas leaves the base of the tower and the undissolved gas leaves the top of the tower. The liquid may be recycled until it will dissolve no more gas, or it may be "stripped" of its dissolved gas and used again.

Bubble Plate or Sieve Plate Columns

These are sometimes used instead of a packed column and are often more economical to operate, but the presence of solids in the liquid may tend to block the fine holes in the sieve plate column.

Spray Towers

This type is a short, unpacked tower into which liquid is injected at the top as a very fine spray. They are not used frequently but are employed in air-conditioning plants and as vent condensers.

Jet Scrubbers

The absorbing liquid is forced under pressure through a nozzle and into an orifice, then dispersed in a chamber where the gas is sucked in and absorbed. Caustic soda solution recycled continuously may be used to remove acid gases: if water is used it is discharged to an effluent treatment plant.

CALCINATION

This unit operation may be defined as the heating of materials to a high temperature in rotary calciners or kilns to convert metals to their oxides, carbonates to their oxides, burn off unwanted organic substances, and for drying. In the latter case it is used particularly for drying substances which contain water of crystalization, which, since it is chemically bonded to the substance, is more difficult to remove than from those which are simply "water-wetted".

The *rotary calciner* is similar to a rotary drier, except that it is specially constructed in stainless steel or special metals lined with refractory materials which can withstand the higher temperatures required. The direct-heating type consists of an inclined metal cylinder lined with a refractory material, down which the crushed material being treated passes continuously from screw conveyors or hoppers and is discharged at the lower end. The lower end is connected to a furnace heated by burning gases; for example, producer gas or a solid fuel may be mixed with the material being changed to the calciner itself. It is sometimes necessary to fit gas absorbers (see p. 61) to avoid passing noxious gases to the atmosphere, or cooling chambers at the discharge end using sprays of cooling water. *Indirect-heat calciners* employ the same principle, but use a stainless steel cylinder rotating within a stationary cylinder lined with refractory material. The fuel gas is burned between the walls of the inner and outer cylinders, and the material fed down the inner cylinder. The inner cylinder extends beyond the ends of the outer cylinder to accommodate the bearings, driving machinery, feeders, etc. Sometimes ores break down to a powder by the calcination process and this eliminates the need to crush the material.

CRYSTALLIZATION

Solids occur in the form of crystals when the atoms and molecules within the bulk are arranged in an ordered manner to form a definite geometrical pattern or shape. This arrangement gives rise to a pure form of the solid since impurities will not fit into the spaces and are rejected. The formation of crystals by a crystallization process is, therefore, often used to produce pure chemicals. Furthermore, crystallization improves the appearance of chemicals, makes handling, and often drying, easier.

Crystallization may be defined as the removal of a solid from a solution by increasing its concentration above its saturation point, that is by becoming "supersaturated". This is achieved by cooling, evaporating, or both.

The simplest method of crystallization is the addition of a solid to a hot liquid (the solvent) until no more will dissolve. The saturated solution thus produced is allowed to stand and cool, perhaps over a few days, when crystals are deposited on the bottom of the tank, leaving a *mother liquor*. The mother liquor is separated by filtration, and the crystals freed of mother liquor wetting their surfaces by careful washing with cooled, pure solvent. The crystals thus obtained are called the *first crop*. To increase the yield, the mother liquor is sometimes evaporated to a smaller volume and the crystallization operation repeated. In this case the liquor when separated from the crystals—the *second crop*—is often called the *grandmother* (or *granny*) liquor.

Decolourizing charcoal is often added to the hot, saturated solution prior to cooling. After agitating the suspension for a short time the charcoal is removed by filtration and the crystallization is completed in the normal manner in another vessel. If charcoal is not used, the hot solution is normally passed through a *polishing filter* to remove small quantities of undissolved impurities or dirt and obtain a clear, bright solution (p. 47).

Sometimes it is necessary to add a small quantity of crystals to the hot solution to start the crystallization process; this is known as *seeding*. The same effect may be achieved by gently scraping the inside wall of the tank, by agitation, or simply by contamination with atmospheric dust.

Agitated batch crystallizers are tanks, often with a tapered bottom, fitted with a suitable agitator and cooled by circulating cooling water or brine through an outer jacket or an immersed coil. This method (i) is quicker, by virtue of the forced cooling and the better heat transfer achieved by the agitation; (ii) produces crystals of similar size by keeping the temperature of the mother liquor more uniform and preventing the small crystals from settling to the bottom of the tank; (iii) prevents the crystals from sticking together.

Simple evaporators used for crystallization may be shallow, open pans fitted with a steam jacket. The crystals deposited by the evaporation of the solvent (usually water) are prevented from sticking together or to the pan walls by slow-moving rakes.

Vacuum crystallizers are closed, lagged vessels to which vacuum may be applied to a hot, concentrated solution to induce cooling by evaporation.

Continuous crystallizers employ the recirculation of cooled, supersaturated solution through a loosely packed bed or suspension of crystals when crystal growth occurs. The thick magma or

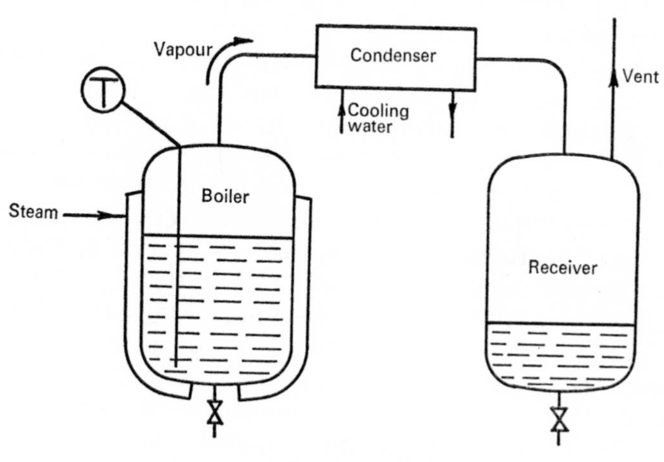

FIG. 4.2. Simple distillation plant.

slurry is continuously siphoned or pumped off and filtered. The mother liquor may be saturated again and returned to the system.

DISTILLATION

The term *simple distillation* generally refers to the separation of a mixture of two or more liquids by boiling, to produce a vapour of the more volatile (lower boiling point) component and the condensation of this vapour back to a liquid which is collected in a receiving vessel (Fig. 4.2).

This unit operation depends on the fact that liquids have different vapour pressures at a given temperature. When a liquid mixture is volatilized the vapour produced is richer in the more volatile component and the liquid remaining in the boiler is richer in the less volatile component. Hence a partial separation of the liquids is achieved. If the condensed vapour—the *distillate*—is heated again, the vapour produced will be even richer in the more volatile component. This process of redistillation can be repeated many times until the separation of the liquids is nearly complete, but this is a lengthy operation and may be achieved more conveniently by fractional distillation.

Fractional distillation or *rectification* is a combination of redistillation and partial condensation of the vapour and employs a fractionating column. Part of the condensed vapour is returned to the boiler down the fractionating column where the liquid is brought into contact with the vapour rising up the column on its way to the condenser. This results in the less volatile components of the liquid mixture in the column becoming enriched, whilst the vapour leaving the top of the column is enriched in the more volatile components. The liquid condensate being returned to the boiler is called *reflux*, and the *reflux ratio* is the number of gallons (or pounds) of liquid sent back per gallon (or pound) of the liquid collected in the receiver. Generally, the higher the reflux ratio the greater the efficiency of separation. Columns may be filled with a packing material such as Raschig rings, Berl saddles, or small glass cylinders, or they may contain a series of perforated bubble-cap plates or sieve plates.

Continuous distillation is achieved by returning the liquid condensate to a point approximately midway up the fractionating column, where the liquid in the column is about the same composition as the returning liquid (reflux). The relatively pure vapour leaving the top of the column may be taken off continuously, leaving only the unwanted "still bottoms" in the boilers. The section of the column above the inlet is called the rectifying section, and that below is called the stripping section.

Steam distillation is a special method employed for distilling liquids with a high boiling point or those that decompose when heated to their boiling point. This is carried out by heating the liquid in the presence of water or injecting live steam into it through a pipe. The operation usually involves two immiscible liquids (for example aniline and water). In this case, the vapour mixture condenses to give two liquid layers which may be separated in the receiver (see Solvent Extraction). Another method employed for distilling heat-sensitive compounds is *vacuum distillation*: the pressure in the system is reduced to lower the boiling point of the liquid to be distilled.

DRYING

This unit operation is used in most chemical processes as chemicals are generally required for use in the dry state and, in any case, it is costly to transport the water or solvent contained in "wet" materials. Drying may be broadly defined as the removal of water or a solvent from solids, liquids, or gases. The term is usually applied to the removal of relatively small quantities of water or solvent, whereas evaporation (see p. 96) involves larger quantities.

The removal of water from gases may be achieved by passing a stream of the "wet" gas through special chemical drying agents (such as concentrated sulphuric acid, anhydrous calcium chloride, or silica gel), by rapid cooling or by physical methods. Removal of water from solvents is effected by azeotropic distillation or evaporation, freezing, or by the addition of chemical drying agents. The drying of solids, which is the most important application of this operation, is achieved in a drier in two steps; the supplying of heat and the removal of vapour. Types and design of driers are discussed in Chapter 3.

Driers may be classified by the method of transferring the heat to the wet solid. *Direct* driers employ a stream of hot gas passing over or through the wet solid which vaporizes the liquid and carries it away, whereas *indirect* driers rely on the transfer of heat through a metal wall. The choice of drier depends upon many factors which include:

(i) Water or solvent content of the wet solid (percentage by weight).

(ii) Temperature required.
(iii) Form of material, for example powder, lumps or slurry.
(iv) Nature of the solid and solvent to be removed, e.g. poisonous, corrosive, or flammable.
(v) Drying time (or drying cycle).
(vi) Bulk density of wet and dry solid.

Many solids are sensitive to high temperatures and may discolour or decompose if overheated: drying at lower temperatures is usually achieved by the use of *vacuum driers*. Overheating may also cause some solids to melt and form glass-like beads, or a molten mass, which sticks to the supporting medium. Particular care must therefore be taken with solids with low melting points (these are usually organic substances) to control the drier temperature.

A special method of vacuum drying for chemicals which are very sensitive to heat is called *freeze-drying*. An aqueous (water) solution of the material is frozen in shallow trays and placed in a chamber which is then evacuated to a high vacuum. If the vacuum is kept high enough to maintain the water-vapour pressure in the chamber below the vapour pressure of the ice itself, the ice will pass directly into the vapour state without melting, leaving the dry material on the trays.

Rotary and tumbler driers are most suitable for drying solids which do not conduct heat very well or are very dusty or hazardous, since this type keep physical contact and handling to a minimum. It is necessary, however, that the solids break down easily into a powder as the moisture or solvent is removed, otherwise the material tends to form a hard cake on the walls, or form into balls which will not dry easily. The load of material in this type of drier should preferably be about a half of the internal volume to attain maximum efficiency.

Tray driers are simple to operate and commonly used, but involve considerable manual handling of material. It is important that the wet solid should be spread thinly and evenly on the trays, usually to a depth of no more than half an inch. Too thick a layer of solid will result in a long drying cycle, an uneven rate of drying throughout the bulk, and the tendency to form lumps. It is desirable to check the internal temperature of the drier frequently, particularly those which do not have automatic temperature control. It is often advantageous to turn the material over and carry out regular visual inspections during the drying cycle. Any visible change in its appearance or form should be reported to the supervisor immediately. At the end of the normal drying cycle a sample of the dry solid is taken and submitted to the analytical laboratory for a moisture determination (see Chapter 2). When unloading trays from racks it is good practice to start removing them from the bottom, since this avoids the risk of dislodged dirt or dust falling on the contents of the tray below. The principle is applied in reverse when loading trays on racks (the first tray is placed on the top of the rack).

The discharge of driers often gives rise to dust in the atmosphere which may be harmful. Dusts may be poisonous, corrosive, dermatitic, and, in some cases, explosive. The chemical operator must keep dust to a minimum and ensure he is adequately protected by covering exposed skin and wearing a dust mask.

EVAPORATION

Evaporation is the separation of one liquid from another or from a solution or suspension of solid particles by changing the liquid (often water) to the vapour state. This operation differs from that of drying in that relatively large amounts of liquid are involved, thus evaporation often precedes drying in chemical processing. Distillation is concerned with the vapour itself and the separation of its components, whereas evaporation processes are usually concerned with the material which remains in the evaporator.

Evaporation is achieved by the application of heat, which may be supplied by the sun, fire, steam, hot water, electricity, hot oil, or special materials. This heat is used to (i) heat the compound to its boiling point, (ii) supply the latent heat of vaporization (which is the amount of heat required to change a compound from liquid to vapour at its boiling point). Since the supply of heat is expensive, the design and operation of evaporators is concerned primarily with its efficient utilization. For example, the heat of the vapour may be utilized to heat another vessel (multiple effect evaporators). Heat transfer depends on the temperature difference between the materials supplying and receiving the heat, the area over which it is being transferred, and the material of the heating surface.

Most evaporators employ steam-heated, tubular heating surfaces, over which the liquid to be vaporized is circulated. The efficiency of an evaporator may be reduced considerably by the formation of scale on the heating surface, which may be removed by chemical or mechanical means. Another problem is the formation

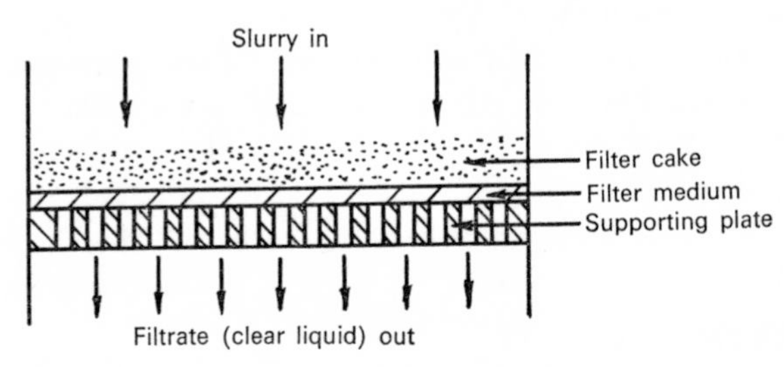

FIG. 4.3. A simple filter.

of foam which may seriously restrict the capacity of an evaporator, but this may be controlled by the use of special chemical defoaming agents or gas or air jets.

FILTRATION

This unit operation covers a wide range of mechanical methods designed to separate solids from liquids or gases. In a simple filter (Fig. 4.3) the mixture of solid and liquid, or solid and gas, is pressed against one side of a porous bed called the *filter medium*, allowing only the liquid or gas to pass, leaving the solid in the form of a cake on the surface of the filter medium.

The liquid containing the suspension of solid particles to be separated is called the *slurry*, and the clear liquor obtained by the separation, the *filtrate*. The wet solid retained by the filter medium is known as the *filter cake*. This cake must be washed with pure water or solvent to remove the residual filtrate adhering to it; this is called the *filter wash*.

There is a resistance to flow of liquid passing through a filter medium which causes a pressure drop across the filter. This resistance increases as the solid cake builds up on the filter medium and it is necessary, therefore, to apply a force to maintain the filtration at a steady rate. The nature of this force provides a convenient means of classifying the many types of filter. Thus, we have *gravity* filters, *vacuum* filters, *pressure* filters, and *centrifugal* filters (see Chapter 3).

Gravity filters are the least efficient since they rely only on the weight of the liquid itself to cause it to flow through the filter medium. Hence they are used only for filtering solids with a large particle size, requiring a filter medium of relatively large pores to prevent their passage through it, and therefore a low resistance to flow. On the other hand pressure filters may be operated with slurry inlet pressures of up to 60 psi to handle finely divided solids. Therefore, the choice of medium depends on the solid to be filtered and the type of filter to be used. A further consideration is the chemical nature of the liquid, for example whether it is acid or alkaline, corrosive, or capable of dissolving the filter medium. A wide variety of filter media are now available in different pore sizes, each offering resistance to particular chemical attack. They include cotton, jute, wool, paper, cellulose–asbestos, woven metal cloth, glass cloth, rubber cloth, and a range of synthetic materials, such as Terylene, Nylon, and polypropylene. When fitting a filter medium, care must be taken to place it perfectly flat on the supporting plate, frame, or wall, and in a symmetrical position so that the edges may be properly sealed to prevent solids by-passing it. Naturally, a torn or holed cloth is useless and must be replaced.

When a slurry contains finely divided solids, or is slimy or viscous, it may be necessary to use a *filter-aid*. Filter-aids are light, powdered materials (such as decolorizing charcoal, magnesia, or diatomaceous earth) which may be added to the slurry or spread over the filter medium, as a *pre-coat*, to prevent clogging of the pores and speed filtration.

The operating steps in a filtration procedure usually consist of the following:

1. *Setting Up*
   (a) Check that plates, frames, or other supporting media are clean.
   (b) Select filter medium, inspect, and fit in correct position on supports.
   (c) Check that components are assembled in the proper order.
   (d) Seal off the whole assembly.
   (e) Set all associated valves in correct positions.
   (f) Apply heat to jacket as necessary.

2. *Loading*
   (a) Start slurry feed to inlet side of filter.
   (b) Adjust internal pressure, vacuum, or speed of rotation to that required.
   (c) Regulate rate of slurry feed as necessary.
   (d) Check that filtrate is free from solid (take sample if necessary).
   (e) Check that filtrate flows freely and evenly.
   (f) When filtrate ceases to flow, or cake is required depth, stop feeding slurry to filter.

3. *Washing*
   (a) Continue to rotate (centrifuge), or apply vacuum or compressed air or gas until filtrate flow stops.
   (b) Feed pure wash liquor to filter inlet and repeat (a).

   The whole operation may be repeated as required.

4. *Discharge (manual)*
   (a) Open up filter by releasing clamps or opening lid.
   (b) Remove solid by scraping from the filter medium (for plate and frame filters, clear one frame at a time).
   (c) Readjust filter medium or wash down filter.
   (d) Take representative sample of filter cake.

PRECIPITATION

A precipitation process involves the formation of a *precipitate*, which is a suspension of particles of a solid in a liquid. If a chemical reaction takes place in a solvent to form a product which will not dissolve in that solvent, then the product will be rejected as a solid precipitate. Another method of forming a precipitate is by the addition of a concentrated solution of a dissolved solid to a large quantity of a pure liquid in which the solid will not dissolve. This process is used extensively to isolate chemical products or waste material from process liquor.

Precipitation should not be confused with crystallization (see p. 63). A precipitate is produced instantaneously, its particles are amorphous (that is they have no particular shape or form) and they do not "grow" in size like crystals. Precipitated solids are often of a lesser purity than crystals since sometimes spongy aggregates of particles are formed which trap (occlude) liquid or solid impurities within the bulk. For this reason, vessels used for precipitation are usually provided with very efficient means of agitation to ensure that the particles of the precipitate are properly dispersed in the liquid. In most cases, only limited cooling is necessary, and sometimes it is not required at all.

SIZE REDUCTION AND SIZE SEPARATION

Size reduction may be described in many terms, such as crushing, grinding, cutting, shearing, comminution, breaking, disintegration, or pulverization. The choice of term depends on the machine used to reduce the size of the particles, the method it employs, and the degree of fineness required. This is a very common unit operation since many materials cannot be used conveniently unless they are reduced in particle size, or made a convenient shape. Materials may be available in the form of boulders, lumps, crystals, or agglomerates, but are often required for use as fine powders. For example, a crystalline material may be milled to a fine powder to enable it to be compacted into a tablet for effective medicinal use. In some cases, material is required in the form of a powder to enable it to be dissolved readily in a solvent, to chemically react, or simply to flow freely in processing equipment. In all cases size reduction brings about an increase in surface area of the solid and it is this effect for which the operation is most utilized.

The various types of mills and crushers used for this operation are described in Chapter 3. Size reduction may be achieved in several stages using different devices: for example large boulders or rocks may be reduced to lumps or pebbles using a jaw crusher, then reduced to a coarse powder using a hammer mill, and finally, to a very fine powder using a micropulverizer.

Size separation is a process which often follows size reduction: its purpose is to separate and classify the solid particles according to their size. The main methods are screening, magnetic separation, and mechanical classifying. Magnetic separation is limited to iron and similar metals, and is often used to separate iron impurities from other materials to eliminate damage to chemical plant or discoloration of the products.

Classifiers work on the principle that the rate at which solid particles settle in a liquid depends on the particle size. The larger particles may be drawn off at the bottom and the smaller particles as an overflow at the top. A device of this type, used for the classification of very fine particles, is called an elutriator, and the operation of a series of such devices, *elutriation*. A suspension of about 10 per cent solids is fed into the side of a column against an upward current of water when smaller particles flow upwards and overflow into another vessel and larger particles settle out at the bottom.

Screens are widely used for the separation of wet or dry solid particles. They consist of metal bars, perforated, holed, or slotted metal sheets, or fine wire mesh. The material to be separated passes over the screen which is shaken or vibrated, manually or mechanically, to allow all the smaller particles to fall through the mesh. The mesh is given a number according to the size of the hole or aperture, and for wire-mesh screens the number indicates the number of holes per inch of the screen surface. Standard scales, which take into account the width of the wire itself, are given by the British

Standards Institution (B.S.I.), the Institute of Mining and Metallurgy (I.M.M.), the United States and Tyler Standards.

SOLVENT EXTRACTION

There are two types of solvent extraction: solid–liquid and liquid–liquid extraction. The former, often called *leaching*, is the process of removing a substance from a mixture of solids by mixing with a liquid (the solvent) in which the substance it is required to separate is dissolved. Liquid–liquid extraction is based on the same principle, but in this case a solution of a dissolved substance is intimately mixed with another liquid (the solvent). If the two liquids are immiscible, two separate liquid layers are formed when the mixture is allowed to settle, the dissolved substance being concentrated in the second liquid. The liquid now containing the dissolved substance is called the rich layer or *extract* and the remaining spent layer is called the *raffinate*. In many cases one of the solvents used is water, and these are termed *aqueous extractions*. Solvents commonly used in conjunction with water are chloroform, carbon tetrachloride, ethylene dichloride, benzene, toluene, xylene, and petroleum ethers.

A simple batch, liquid–liquid extraction is illustrated in Fig. 4.4.

If the liquids are partially soluble (that is, part of each liquid dissolves in the other), it may be necessary to perform a second extraction when the raffinate is mixed again with fresh solvent. It is convenient if the initial solvent chosen is less dense than the added pure solvent, because it forms an upper raffinate layer. In this way mixing of the solvents may be assisted, and, should a second extraction be required, it may be carried out directly in the same extraction vessel after running off the first extract.

A solvent extraction process is usually followed by distillation to remove the solvent from the extract and render the solvent available for re-use.

A simple flowsheet for a *continuous, single extraction* process with solvent recovery is shown in Fig. 4.5.

Mixers achieve intimate contact between the two liquids by

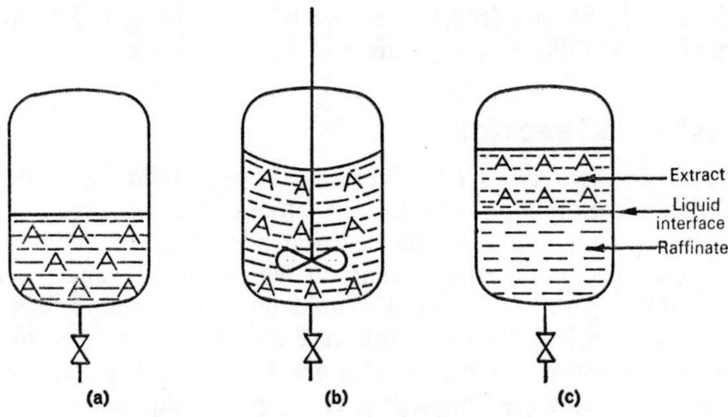

FIG. 4.4. (a) Solvent containing dissolved substance A.
(b) Second solvent added and mixed by agitation.
(c) Separation of two liquid layers on settling. Raffinate is run off.

mechanical or air agitation or jet mixing (in which one liquid is sprayed through a nozzle into the other liquid) or by external recirculating pumps.

Multi-stage extraction is used to achieve a higher efficiency of separation, in which the product is almost completely removed from the raffinate. The solvent is split up into several portions

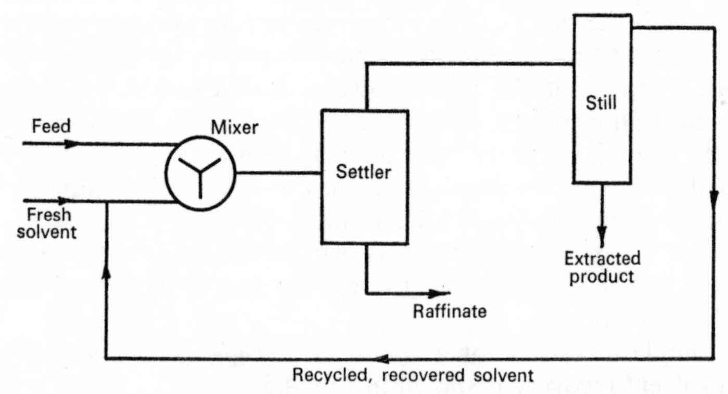

FIG. 4.5. Continuous, single extraction.

and fed to a series of mixers and settlers. The disadvantage of this method is the need to use large volumes of solvent. A more complicated system, called *countercurrent, multi-stage extraction* uses a series of mixers and settlers arranged as before, but the feed liquid and pure solvent are passed through the system in opposite directions, that is countercurrently. Continuous countercurrent operation may be carried out by means of spray columns, packed columns (similar to those used in distillation), plate columns, or, sometimes, bubble cap columns.

Similar methods are employed in solid–liquid extraction—all designed to bring about intimate contact between the solid and pure solvent. Leaching systems, as they are often called, may employ batchwise, single and multiple, continuous, and countercurrent extraction.

CHAPTER 5

Measurements

IN ANY chemical process there are several variable quantities which must be measured to effect proper control and thus ensure safe and efficient operation. Quantities frequently measured are time, temperature, pressure, volume, weight, density, and electrical current, voltage, and resistance. Measuring devices, or instruments, may be of the indicating type, where the instantaneous value is shown, or they may be of the recording type, when the value is recorded continuously on a chart providing a permanent record. A clock, for example, is an indicating instrument for the measurement of time. Nowadays, instruments are often grouped together on an instrument panel or in a control room adjoining the plant.

UNITS

A quantity is a meaningless number unless it is expressed in terms of units. When a measured quantity is recorded for process control purposes, it is misleading, and possibly dangerous, not to state the units in which it is measured. For example, a volume measurement would never be recorded as, say, 6, but as 6 *gallons*, 6 *litres*, or 6 *cubic feet*.

In order to compare the magnitude of quantities—whether physical, electrical, or mechanical—their basic units must be defined. They may be basic (or absolute) units, such as metres (length) or kilogrammes (mass or weight), or derived units, such as volts (voltage) or watts (power). An instrument is marked off in divisions as points on a scale, each division representing one unit of the particular quantity it measures; this is called its *calibration.* The units in which the instrument is thus calibrated are stated on the dial. The divisions on the scale may be divided into subdivisions for taking readings with greater accuracy. On a metric scale, these subdivisions represent one-tenth or 0·1 of a whole unit.

An extension and refinement of the metric system, called S.I. units (Système International d'Unités) is being adopted gradually in Great Britain. It has six basic units: metre (length), kilogramme (mass), second (time), ampere (electric current), degree Kelvin (temperature) and candela (luminous intensity). Multiples are normally expressed in steps of 1000 and fractions to $\frac{1}{1000}$. Adoption of this system throughout the world will avoid confusion and time-wasting conversions between the several different systems now used (Table 3).

MEASUREMENT OF TEMPERATURE

Temperature may be defined simply as the degree of "hotness" or "coldness" of a material or of the atmosphere. Heat is a form of energy and temperature is a measure of the level of this energy. *Thermometers* make use of a number of different properties of substances which change with temperature: for example, the expansion of a liquid or the variation of electrical resistance.

The basic temperature scale is the absolute or Kelvin scale. The *absolute zero of temperature* is the lowest temperature theoretically possible, and is so low that the scale cannot be used conveniently in chemical processing. Therefore, in practice, we use the Centigrade and Fahrenheit scales. These scales utilize fixed points, which are normally the freezing point and boiling point of pure water. Thus we have:

	Temperature scale		
	Absolute	Centigrade	Fahrenheit
Freezing point of water	273°A (or °K)	0°C	32°F
Boiling point of water	373°A (or °K)	100°C	212°F
Range	100°A (or °K)	100°C	180°F

TABLE 3

Quantity	Units	Symbol or abbreviation
Temperature	Degrees: Centigrade	°C
	Fahrenheit	°F
	Absolute (or Kelvin)	°A (or °K)
Pressure	Pounds per square inch	psi
	Kilogrammes per square centimetre	kg/cm² (kg.cm⁻²)
Density	Grammes per cubic centimetre	g/cc (g.cc⁻¹)
Flow	Cubic feet per minute	ft³/min (ft³.min⁻¹)
	Gallons per minute	gal/min (gal.min⁻¹)
	Cubic metres per hour	m³/hr (m³. hr⁻¹)
Weight	Kilogrammes	kg
	Pounds	lb
Heat	Calories	cal
	British thermal units	B.T.U. (or Btu)
Volume	Litres	l.
	Gallons	gal
	Cubic feet	ft³
	Cubic centimetres	cc
Current	Amperes	A (amp)
Voltage	Volts	V
Resistance	Ohms	Ω

To convert degrees Centigrade to degrees Fahrenheit:

$$(°C \times \tfrac{9}{5}) + 32 = °F.$$

To convert degrees Fahrenheit to degrees Centigrade:

$$(°F - 32) \times \tfrac{5}{9} = °C.$$

The Centigrade scale is used mainly in chemical processing, whereas the Fahrenheit scale is often used for plant-heating systems and other services.

As the range of temperature in chemical processes may vary from well below 0° C to several hundred degrees centigrade, a wide range of measuring techniques have been evolved. The common *mercury-in-glass thermometer*, which depends on the expansion of the liquid when heated, is not often used in industry. Thermometers of this type, with steel containers filled with mercury or other suitable liquid (such as alcohol for low-temperature measurement), may be conveniently used, particularly if fitted with some means of remote measurement.

The most important type of thermometer, the *thermocouple*, uses the electrical effects of temperature. It consists of two wires of dissimilar metals, joined at the ends and connected to a current-measuring instrument. When one of the two junctions is heated an electrical current flows between them and the measuring instrument is calibrated to read the temperature of the hot junction directly. Common thermocouples are copper–constantan and chromel–alumel.

Instruments used for temperatures higher than can conveniently be measured by thermometers are called *pyrometers*. Typical pyrometers are as follows:

Electrical resistance pyrometers	(up to 600° C).
Thermo-electric (thermocouple) pyrometers	(up to 1450° C).
Optical pyrometers	(over 600° C).
Radiation pyrometers	(over 500° C).

The *electrical resistance pyrometer* depends on the principle that the electrical resistance of a conductor changes with temperature; the instrument is designed to measure these changes and is calibrated in units of temperature. The circuit used is the Wheatstone Bridge circuit and the thermometer bulb usually contains a platinum element.

Optical pyrometers are suitable for measuring temperatures up to 3000° C and are usually of the "disappearing filament" type. The heated object is viewed through a telescope which is fitted with an electric lamp. The brightness of the lamp filament is adjusted until its temperature equals that of the object, when the lamp filament "disappears". At this point the temperature of the filament, and hence the object, can be obtained from calibrated scales.

Radiation pyrometers use the principle that the wavelength of the radiation from a hot source depends upon its temperature. These pyrometers have the advantage that there is no physical contact between the instrument and the source of heat.

MEASUREMENT OF PRESSURE AND VACUUM

Pressure is defined as a force applied to unit area of a surface. It is measured in pounds per square inch (psi) or kilogrammes per square centimetre (kg/cm^2 or $kg.cm^{-2}$). Atmospheric pressure is the pressure exerted by the atmosphere on the earth and varies with altitude and weather conditions, but under standard conditions it is approximately 14·7 psi. Vacuum is a pressure which is lower than atmospheric pressure and pressures approaching zero represent a "high vacuum". Thus, the lower the pressure the higher the vacuum. Vacuum gauges are calibrated in inches or millimetres of mercury, torr, or percentage vacuum: the maximum readings being 30, 760, 760, and 100 respectively.

Most pressure gauges measure the pressure above atmospheric pressure and this is called the *gauge pressure*. Some gauges will indicate both pressure and vacuum on the same dial and are called compound gauges.

The simplest method of pressure measurement is by means of of a *U-tube manometer* which consists of a glass tube bent in the shape of a U and filled with a liquid, such as mercury or water: it is connected at one end to the vessel under pressure. This device is most suitable for measurement of low pressures (Fig. 5.1).

The most widely used pressure instrument in the chemical industry is the *Bourdon gauge*, which may be used for higher pressures (Fig. 5.2). A common type consists of a metal tube, closed at one end, and bent in the arc of a circle, which tends to increase its radius (straighten out) on the application of pressure within

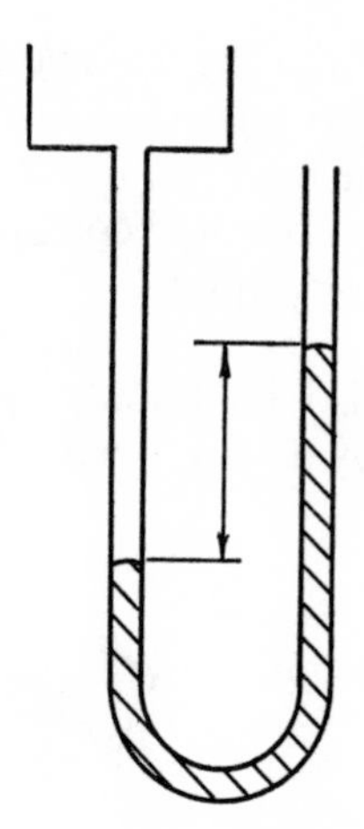

FIG. 5.1. U-Tube manometer.

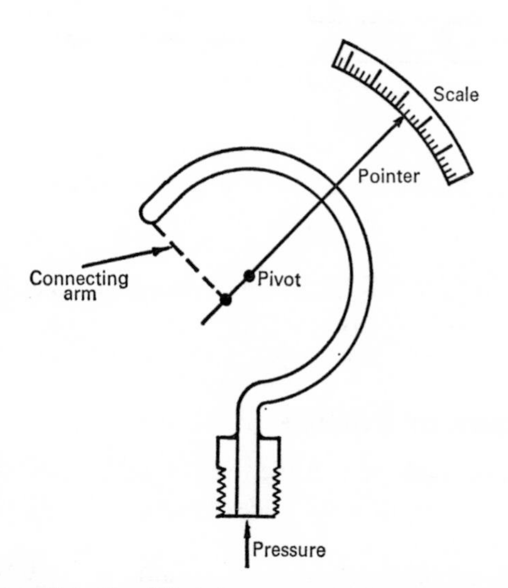

FIG. 5.2. Simplified diagram of Bourdon gauge.

the tube. A connecting link from the free end of the tube is joined to a pivoted indicating pointer placed in the centre of a dial.

The Bourdon gauge may be modified to measure the pressure of corrosive liquids or gases, which would corrode the metal tube, by sealing it off with a flexible diaphragm.

Diaphragm gauges are also commonly used, particularly as vacuum gauges. They employ a thin, flexible diaphragm which is

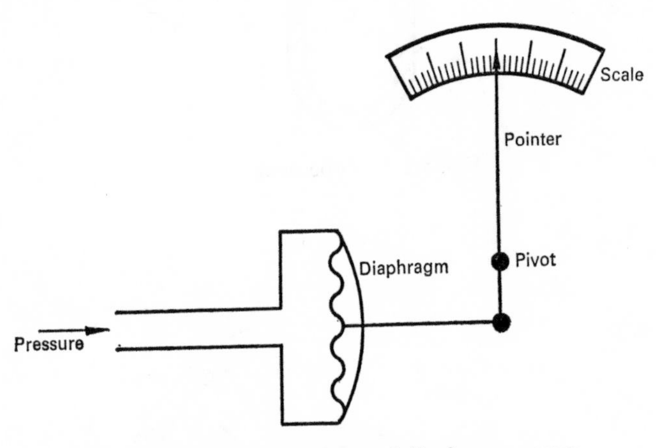

FIG. 5.3. Simplified diagram of diaphragm gauge.

pushed outwards when pressure is applied: the movement is shown by an indicating pointer (Fig. 5.3).

MEASUREMENT OF VOLUME

The common units of volume are the *gallon* and the *litre*. The litre, which is the volume of 1 kg of pure water at 15° C (60° F), is the metric unit.

The British imperial gallon is the volume of 10 lb of pure water at 15° C (60° F) and is sometimes called a metric gallon. The

United States gallon is a smaller volume and is equal to 0·8327 of an imperial gallon.

1 imperial gallon	= 4·546 litres.
1 imperial gallon	= 1·2009 U.S. gallons.
1 U.S. gallon	= 0·8327 imperial gallons.
1 litre	= 0·22 imperial gallons.

The simplest instrument for the measurement of the volume of a liquid in a vessel is the *dipstick* or *gauge stick*. The dipstick is a straight rod, calibrated along its length, which is immersed vertically in the liquid so that its lower end touches the lowest point in the

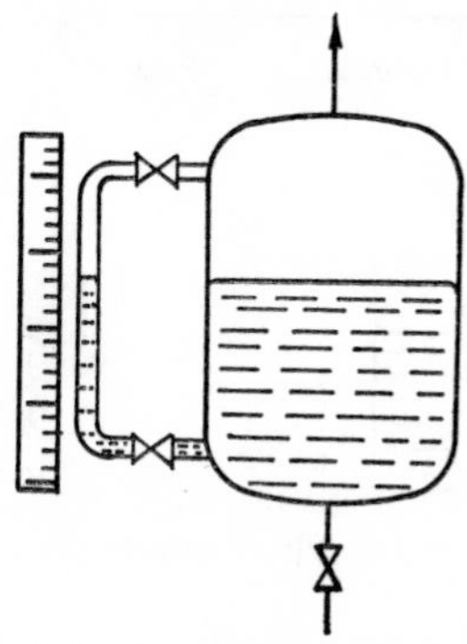

FIG. 5.4. Vessel fitted with gauge glass.

essel. On raising the dipstick the volume of liquid in the vessel may e read from the mark coinciding with the liquid level. A dipstick an only be used in the vessel for which it has been calibrated.

A common device for measuring the volume of liquid in vessels is he *gauge* (*or sight*) *glass*. The glass is calibrated by marking the olume corresponding to the depth of liquid in the vessel. Valves are sually fitted at the upper and lower ends of the glass to prevent he loss of the contents of the vessel if the glass is accidentally roken. These valves should be closed when the reading has been aken. The vessel should be vented to the atmosphere when a eading is being taken since a vessel under pressure or vacuum may ive a false reading in the gauge glass (Fig. 5.4).

Another device which depends on liquid level is the *float gauge*. In gauges of this type, a float is connected to a counterweight outside the vessel by means of a chain. The weight moves up and down against a scale as the float rises or falls (Fig. 5.5).

DETERMINATION OF WEIGHT

Weighing materials is usually the most common measurement in chemical processing. Unfortunately, it is also the most common source of error.

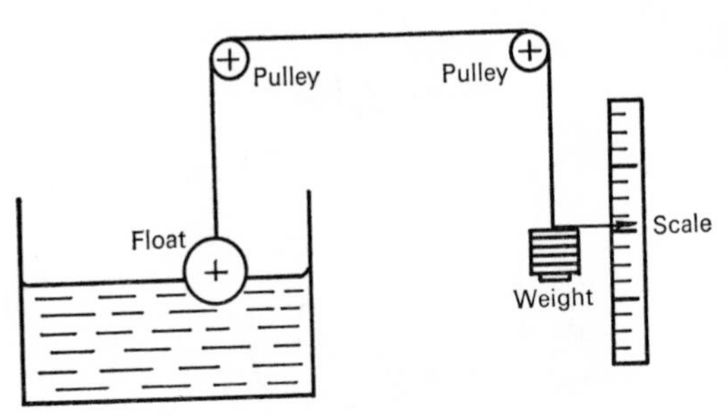

FIG. 5.5. Float gauge.

The units of weight or mass are the pound and the kilogramme, which are related as follows:

1 lb = 453·592 g.
1 kg = 1000 g = 2·2046 lb.

Platform scales used in chemical plant are usually of the *spring* or the *steel yard* type, with or without a tare bar. The latter enables prior adjustment to be made for the weight of the container. Some general points of weighing technique, to ensure accuracy, are given below:

(i) Check that scales are in a level position and cannot slide.
(ii) Check that the pointer is in the zero position (if not, adjust the tare bar).

(iii) Check the tare weight of the container, including the bungs or lid, *before* filling with material.
(iv) Lower the load *gently* onto the platform (with some scales the arm or dial may be locked beforehand).
(v) When reading a dial, the eyes must be directly in front of the pointer.
(vi) Record the reading carefully, showing gross, tare, and net weights, and their units. Net weight equals gross weight less tare weight.
(vii) Clean the platform and moving parts regularly.

MEASUREMENT OF DENSITY

Density is defined as the weight per unit volume of a substance at a given temperature.

$$\text{Weight} = \text{density} \times \text{volume},$$

$$\text{therefore, density} = \frac{\text{weight}}{\text{volume}}.$$

The common units of density are grammes per cubic centimetre. The density of pure water is 1, thus 1 cubic centimetre weighs 1 gramme.

The specific gravity of a substance is the ratio of its density at a particular temperature to the density of water at 4° C. It is numerically equal to the density in grammes per cubic centimetre at that temperature and, being a ratio, has no units. The specific gravity of liquids may be measured by means of a *hydrometer*, which consists of a graduated glass tube attached to a bulb weighted with mercury or lead shot (Fig. 5.6). The depth of immersion of the tube in the liquid indicates the specific gravity of the liquid. The lower the specific gravity the lower will the hydrometer sink. It is important to remember that the hydrometer must not touch the side of the vessel or it will tend to stick and give a false reading.

MEASUREMENT OF pH

Determination of the pH value of a solution is important for the control of many chemical processes. This value which ranges from 0 to 14, is a measure of the acidity or alkalinity of a solution and is related to the concentration of hydrogen ions on a logarithmic scale. A molar solution of a strong acid has a pH of 0 and the strongest alkali (base) has a pH of 14. The mid-point in the scale,

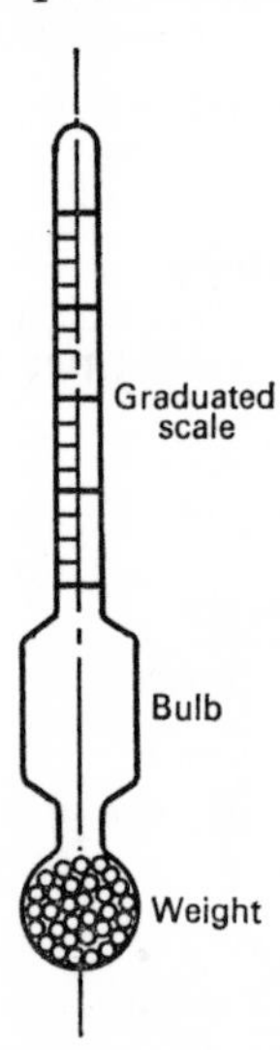

FIG. 5.6. Hydrometer.

pH of 7, is that of a neutral solution or pure water, which is neither acidic nor alkaline.

An approximate measurement of pH may be made by immersing in the solution a strip of indicator paper, which turns a particular colour at a certain pH value. The pH may be determined by comparison of the colour of the indicator paper with a colour chart. These papers are available for measuring pH over a wide or narrow range of values.

Accurate measurement of pH is achieved by the use of electrical pH meters. Potentiometric meters consist of a measuring electrode (hydrogen gas, quinone, glass, or antimony) and a reference electrode or standard electrode (usually calomel) connected to a potentiometer circuit which measures the voltage generated by the electrical cell when the electrodes are immersed in the solution to be tested. These meters may be adapted to measure the pH of a stream of liquid flowing in a pipe. They may also be connected to a recording instrument to give a permanent, continuous record of the pH of a process liquor.

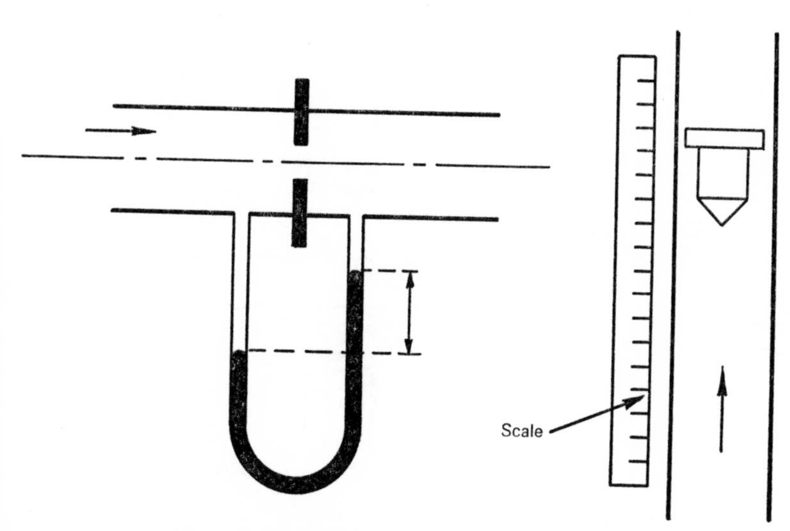

FIG. 5.7. Orifice plate meter and float meter.

MEASUREMENT OF FLOW

Instruments which measure the rate of flow (velocity) of liquids and gases are called *flowmeters;* they may be broadly defined as being mechanical or electronic in operation. Examples of mechanical flowmeters are orifice plate and float meters (Fig. 5.7), venturi meters, and pitot tube meters, all of which depend on a constriction being introduced into the flow stream in order to produce a difference in pressure across the constriction. The rate of flow can then be obtained from the difference in pressure.

The float and taper tube meter (rotameter) consists of a vertical tapered tube in which the float provides a varying orifice between its rim and the side of the tube. In the no-flow position the float falls to the bottom of the tube and seals it off. When liquid is passing, the area available to flow is proportional to the rate of flow and can be measured directly on the tube or remotely by incorporating a magnet in the float and an external follower. The advantage of this type is that the pressure loss is small and almost constant through the whole range.

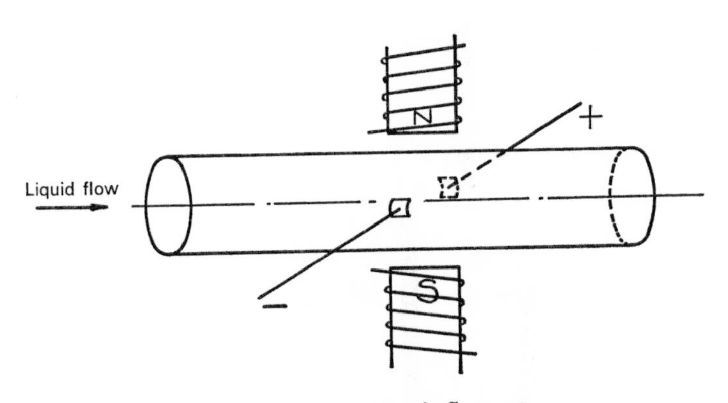

FIG. 5.8. Electromagnetic flowmeter.

More recently, flowmeters have been evolved which make use of electromagnetic induction and ultrasonic principles. These flowmeters have the advantage that there are no moving parts or restrictions in the flow stream and the bore remains the same diameter as the whole system. They are, however, more expensive than the mechanically operated types. With the electromagnetic flowmeter (Fig. 5.8), the flowing liquid acts as a conductor moving through a magnetic field and an electromotive force (voltage) is induced between two metallic electrodes mounted opposite each other in the wall of a tube. This voltage is dependent on the velocity of flow and may be measured by a suitable instrument.

CONTROLLERS

Some chemical processes may be operated automatically, or are dependent on the time taken for certain operations, or they may require certain operations to take place at regular intervals in time. For example, it may be required to monitor or control the time difference between a valve opening, thus allowing liquids or gases to flow, and a second valve closing. In many cases, it is convenient to utilize electrical or air-operated devices to control such operations and these devices are dependent on the opening or closing process to provide a start or stop pulse. A wide range of controllers are used, some of which operate on a simple clock mechanism and others involving electronic counting principles where high accuracy is required.

ELECTRICAL MEASUREMENTS

The International Electrotechnical Commission has defined electrical units as follows:

Ampere. The unit of current in common use, being the unvarying current which, when passed through a solution of silver nitrate in water, will deposit silver at the rate of 1·11800 mg/sec.

Ohm. The unit of resistance in common use, being the resistance offered, at the temperature of melting ice, to an unvarying electric current by a column of mercury 14·4521 g in mass, of uniform cross-sectional area, and 106·300 cm in length.

Volt. The unit of potential difference in common use, being the potential difference, which, when steadily applied to a conductor, the resistance of which is one ohm, will produce a current of 1 A.

Recently the international system of units (S.I. units), which replaces electrostatic and electromagnetic units, was chosen as the legal system in nearly 30 countries and will be adopted in Great Britain in the future.

Ohm's Law. Potential difference, current, and resistance are related by Ohm's Law as follows:

When a steady current flows through a conductor, the potential difference between its ends divided by the current is a constant, provided that the physical condition of the conductor does not change. The constant is termed the resistance of the conductor and is measured in true ohms when the potential is in true volts and the current is in true amperes, or

$$R\,(\text{ohms}) = \frac{V\,(\text{volts})}{I\,(\text{amps})}.$$

Rearranging the equation, we have $V = IR$.

Measurement of Voltage and Current

Instruments which measure voltage and current are called *voltmeters* and *ammeters*. Basically, they are one and the same

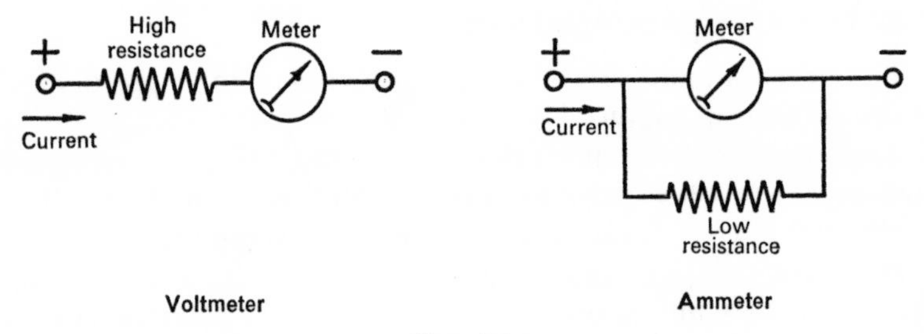

FIG. 5.9.

instrument, since in the case of a voltmeter the reading depends on a current proportional to the voltage to be measured, and for an ammeter, the reading is the current being measured or a fraction of it. This may be illustrated as in Fig. 5.9.

Multi-range instruments have resistors "built in" and the voltage and current ranges are selected by switching in the appropriate resistances. For an instrument to measure alternating currents (a.c.), either the deflection of the meter must be proportional to the mean value of the square of the current or a rectifying circuit may be utilized to convert the alternating current into a direct current.

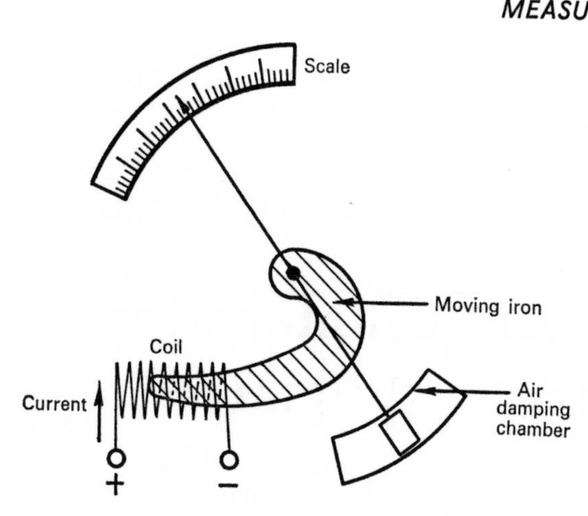

FIG. 5.10. Moving iron instrument.

The main types of instruments are the *moving iron* (d.c. and a.c.) (Fig. 5.10), the *permanent magnet moving coil* (d.c. and a.c.) (Fig. 5.11), and the *moving coil dynamometer* type (a.c.).

The care of instruments is discussed in the next chapter.

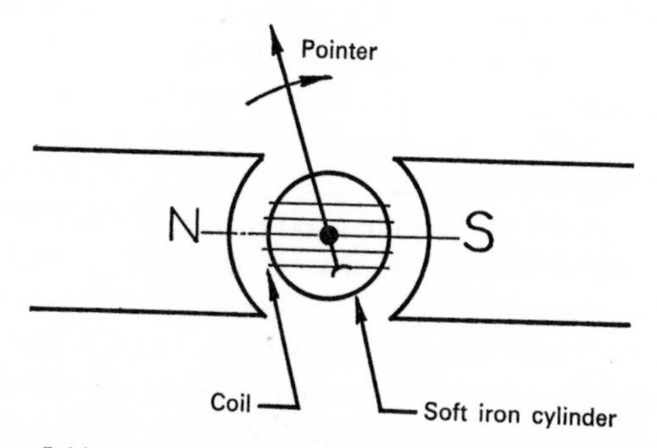

FIG. 5.11. Permanent magnet moving coil instrument.

CHAPTER 6

Care and Maintenance of Chemical Plant

GENERAL PRINCIPLES

THE responsibility for the maintenance of chemical plant is usually, and rightly, vested in an engineering department. Care of plant, however, must be the responsibility of the chemical operator and the extent to which this is exercised is reflected in the amount of maintenance required. For example, a garage is usually expected to maintain a car, but a carefully driven car requires less attention than one which is neglected and driven hard. This means that the careful driver has more use of his car and is less likely to meet with a breakdown. In the same way chemical plant, carefully operated, will require less maintenance and run longer without a breakdown than the same plant operated hard or carelessly.

Care and maintenance, therefore, cover both production and engineering responsibilities: close co-operation and a keen appreciation of each others problems is required to attain greatest efficiency. This implies a conflict of aims by the two departments and there is one sense in which this is so. Ideally, the production department requires the plant to be always in use whilst the engineering department, on the other hand, requires that the plant is not used at all so that it never requires attention. The optimum condition is a carefully operated plant in which the necessary maintenance is carried out, as far as possible, when the plant or item of equipment is not in use.

HOUSEKEEPING

The basic principle of housekeeping is cleanliness and tidiness. It includes routine operations; such as cleaning windows and arranging materials neatly, washing of hands, sweeping up spillages, etc. In the chemical industry this is more important than in most industries because of the nature of the materials handled. Chemicals are often expensive and/or dangerous. The safety aspects of good housekeeping are discussed in Chapter 8.

Dirty-sight glasses or instruments may lead to inaccurate quantities of raw materials being used. Dirty operating handles could result in a hand slipping off a control when it is important to close or open a valve, stop or start a pump or agitator. Chemical operatives are encouraged to be clean and tidy which, in its turn, encourages them to work more conscientiously, efficiently, and safely.

One word of warning, however: cleanliness obtained by washing down with hoses may well remove all dirt from walls, ceiling, windows, and equipment but can also bring many difficulties in its wake. Firstly, drowning everything in water results in all rubbish being washed down the drain, causing blockages and local flooding. Secondly, equipment inside a building is usually not of a weatherproof design and even when it is, it may not resist hosing down. Water-penetrating bearings, gearboxes, etc., will displace or mix with lubricants and lead to premature failures. Water splashed on electrical equipment could cause an immediate spark which is a fire hazard, or blow the fuses by causing a short circuit. It should be remembered that flameproof electrical equipment is NOT weatherproof (although it often looks similar). The use of sweeping brushes, mops, squeegees, dusters, and damp cloths will often give better housekeeping results without the hazards of excessive water.

Maintenance personnel are similarly encouraged by a clean and tidy plant and are more likely to give of their best if the plant has a well cared for appearance. Machinery, no matter how well designed for arduous conditions, will function better if kept clean. A build-up of dust will usually find its way into moving parts of machinery and shorten the life of bearings, gears, belts, chains, etc. The location of faults on dirty plant and equipment is more difficult and usually delayed whilst the item is cleaned.

Finally, it should be stressed that housekeeping is not confined to chemical operators only. Maintenance personnel should leave the plant in a clean and tidy state after a repair. Surplus nuts and bolts, gaskets, packing, and even spanners, have often been found in reactors, blenders, mixers, crushers, etc., resulting in a safety hazard and plant failure. Everyone should play his part in housekeeping; do not leave it to the other man, tidy up as you go.

METHODS OF MAINTENANCE

Broadly, there are two methods of tackling the maintenance of any equipment; do nothing until the equipment fails and then direct all available resources to put matters right or, alternatively, plan inspections and maintenance on the equipment so that failure from fair wear and tear can be anticipated and preventative action taken before failure occurs.

Breakdown Maintenance

Even in factories where planned maintenance is not officially applied the first method of breakdown maintenance is rarely followed completely. When a given item of equipment has failed several times in the same way or after similar intervals of time, someone is sure to suggest that some preventative action should be taken before the time that the next failure is due. This may be as simple as saying that a particular pump, gearbox, or motor is overhauled in the annual shutdown "*because it is known from past experience that it will not last for another year*".

The only argument in favour of breakdown maintenance is that the longest uninterrupted run is obtained from that item of plant, but unfortunately, when it does fail, the period of downtime is unknown; it may be an hour, it may be several hours or days, and this could and does happen during important production runs. In addition to the resultant chaos in the maintenance department there will be idle production operatives and loss of planned production. In the chemical industry this failure could cause a continuous process to shut down and, in some instances, it may take weeks to start up and return to full production. For example, the brickwork in a furnace may require rebuilding resulting from only a short shutdown.

In the case of some batch processes a failure may result in the reaction being incomplete and the batch being ruined, or control of the reaction may be lost. Major breakdowns, when the maintenance department divert all available resources to the rescue, usually means some improvisation or even patching up to meet the urgent need to get back in production as soon as possible. This in turn often leads to another failure shortly afterwards. These sudden demands may have an adverse effect on the morale of all concerned.

Planned Maintenance

The alternative method is to schedule maintenance on a planned basis. This should reduce the loss of production time, reduce the total amount of time spent on maintenance, and, at the same time, even out the work load on the maintenance department.

In its simplest form, planning will take the form of regular (perhaps daily) inspections, greasing, and minor adjustments with a means of reporting any minor defects observed. This may be simply an electric motor running warmer than usual, or the gland of a shaft has been repacked more frequently, or the pressure on a certain gauge is lower or higher than normal. Taken in isolation any single record may appear unimportant, but taken collectively and viewed intelligently, signs of wear or other faults can often be detected. It is for this reason that the chemical operator should report his observations to his supervisor.

Generally, more good than harm is done by regular greasing and checking of oil levels. The regular visitor to a machine will soon be able to detect a strange sound or a change in the running tone and early reporting may save an untimely failure.

More advanced forms of planned maintenance include scheduled stripping down of equipment to enable the internal working parts to be inspected. At these inspections some immediate action in the form of replacement of small parts (bearings or wearing plates) may be required. It may be that the life of a part can be estimated and it can be planned that a replacement be fitted at a time when the equipment is not in use.

A further development of this principle is to plan to replace a worn part in anticipation of its failure. For example, most sparking plugs in car engines are given an estimated life of 10,000 miles by the manufacturers. You may well get 15,000 or 18,000 out of a sparking plug, *but* you are running the risk of a drop in engine performance and even a failure for every mile after 10,000. If you replace on the scheduled plan at every 10,000 miles you will very rarely, if ever, have a failure.

Planned maintenance—in its widest sense—includes all regular scheduled inspections and reports, planning of replacements at regular intervals of time or running hours, or on a "mileage covered" basis, and also the planning of maintenance work shown by scheduled inspection to be required. This will result in the reduction of breakdowns, particularly those due to fair wear and tear. It cannot eliminate failures completely—particularly those which result from neglect, abuse, physical shock, etc. It will, at the same time, highlight those items of equipment which are most susceptible to failure: the effects of such a failure may be reduced by having a replacement available. The most satisfactory method of dealing with equipment that repeatedly fails is to establish its exact duty and investigate the possibility of replacing the equipment with one designed to meet that duty. Frequently, process modifications result in a change of duty for equipment which may be overlooked until repeated failure initiates an investigation.

In every case the aim of planned maintenance is to eliminate breakdowns due to fair wear and tear and to be equipped to deal promptly with any accidental breakdown whatever the cause. The application of these principles to some equipment in general use in chemical plants, together with the care in operation of this equipment, is now considered.

LININGS

The common feature of all lined equipment is the need for support from an outer shell. It is important that this outer shell gives, continuously, the support the lining requires. The lining is, of course, protecting it on one side from the particular corrosive condition, but care must be taken to prevent damage to, or corrosion of, the other face. Precautions must also be taken to prevent any substance finding its way between the lining and the shell. Spillage, or drips from overhead pipes, are frequent causes of damage to outer shells; in particular, pipe joints at the tank nozzles on the top of the vessel should be inspected diligently for leaks which may be difficult to spot (often the leak does not drip but merely trickles down the nozzle, sometimes seeping between the shell and the cover). It is imperative that care be taken to keep clean any drip trays or channels designed to divert or collect any such drips.

Ceramic Linings

The most common ceramic materials used for linings are bricks (including tiles) and glass. Bricks are hard wearing and robust, the weakest point of such a lining being the joint between the bricks or tiles and the sealing of the top or edge of tiling to the vessel or floor. Maintenance must include careful inspection of all joints because if any liquor seeps behind the bricks and is trapped, it can result in sufficient pressure building up to force the bricks away from the floor or walls. All defects must, therefore, be repaired immediately, but care must be taken to ensure that the repair area is dried thoroughly before rejointing is undertaken.

Glass Linings

The method of applying glass to steel (described in Chapter 1), in addition to dictating some limitations on the design of vessels, presents certain inherent weak spots. For example, it is most difficult to glass sharp radii, hence glass in these areas is most prone to damage. Care should be taken to avoid knocking the edges of the flanges where the glass finishes or the bulbous ends of agitator or baffle arms. Other areas susceptible to damage are the inside radii of nozzles, particularly the chargehole, the bottom outlet nozzle, and flanges on the vessel diameter.

There is also a limit of thermal shock that a glass lining will withstand without damage. This means there is a maximum temperature difference between the inner glass face and the vessel shell at which the adhesion will hold. The exact limitation varies with the temperature range, the pH of the liquid, and the type of glass lining; but glass linings are available which withstand a thermal shock of 80°–90° C. Cold water or brine must not, therefore, be applied to the jacket of a vessel containing boiling water, nor must steam be applied while the vessel is holding liquids below 20°C. Live steam jets should never be applied to a glass lining for cleaning purposes.

Considerable damage to glass linings can be caused by permitting acid to drip onto the outside of the vessel; the acid will penetrate through the steel, causing the glass to be blown off the inside of the vessel. Finally, every effort must be made to avoid foreign bodies dropping into vessels, as this almost certainly means some damage which may not be visible. In these incidents it is important to inspect as soon as possible, and repair if necessary, to prevent a spread of the damage which could be very rapid.

Maintenance of glass-lined vessels is primarily through inspection, to detect any damage, however small. The smallest pin-hole which allows acid to make contact with the steel casing will rapidly enlarge. Areas of tight radii, ends of agitator and baffle arms, and radii on each side of gaskets should receive particular attention. *Damaged gaskets must be replaced.* Sheaths when fitted may be split, cut, cracked, or creased and, in some cases, sucked into the vessel. The alignment of the agitator in its nozzle, where there is usually little clearance, must be checked, as any rubbing or knocking here would damage the glass in an area that is most difficult to repair. Externally, regular inspection should be made of the protective paint since acid dripping on the unprotected steel can quickly penetrate the casing and blow off the glass on the inside.

The repair of damaged glass linings is a specialized subject, but many repairs can be readily carried out by skilled tradesmen at a fraction of the cost of reglassing with relatively little interruption to production. There are three main methods of repairing:

(i) Coating the damaged area with several layers of special cement. This repair may have limited life, but is a very effective "temporary" measure.

(ii) Repair by plugging with metallic (for example tantalum) plugs or studs for small areas, and PTFE discs held in place by metal washers for larger areas (up to 6 in. diameter). There are, however, some areas where plugs cannot easily be fitted (such as nozzle areas); in these cases metal sleeves are tailor-made and cemented into position.

(iii) Repairs to damaged nozzles, agitators, and baffles, and large areas of damage to vessel sides can be made with phenolic-bonded asbestos laminate (see Chapter 1). This material can be shaped more readily than metal and can be fixed by furane cement and/or bolting. There is a wide range of standard repair sleeves available for nozzles of various sizes. This material is resistant to most acids and solvents.

Plastic Linings

A wide range of plastics may be used for lining vessels and there are two techniques used. A pre-formed plastic lining can be placed inside a metal casing or frame or it may be bonded or sprayed onto the casing or vessel.

With these vessels, care must be taken to avoid any liquid entering the gap between the vessel and the liner, which might cause the lining to float, bulge, or become distorted, or cause corrosion of the supporting casing. Regular painting will maintain the outer surfaces of both casings and frames.

In the case of bonded or sprayed linings any physical damage to either the lining or casing could puncture the coating or break the bond, exposing the metal to corrosion. Regular inspection will enable damage to be detected early and repairs to be effected. The type of repair will depend on the type of lining and casing but many of the techniques for glass lining repairs can be successfully used.

Rubber Linings

In common with most linings it is important that the adhesion to the metal is maintained. The most likely cause of damage is from falling objects or rodding. Similarly, sharp materials trapped under a lid may easily puncture the rubber and allow acid to attack the metal or the adhesive. Once the acid reaches the metal, corrosion will spread rapidly.

There are many types of rubber used in lining vessels and it is most important to know the limitations of the type being used on a particular vessel. The total range of resistance from rubbers is much wider than the individual resistance of any particular type. Maintenance must include regular inspections and repairs as necessary. The smallest defect should be repaired immediately to prevent the spreading of the damage.

Some damaged areas may reveal a second layer of rubber; sometimes a particular type of rubber will not readily bond to steel and requires another type sandwiched between it and the steel. This inner layer may not have the same resistance as the final layer. Some tanks and vessels are rubber covered on the outside to resist spillage or a very acidic atmosphere. They are more susceptible to physical damage, but can often be repaired immediately while production continues.

PUMPS

The proper selection, use, and maintenance of pumps is perhaps the most important factor which determines the degree of safe and efficient operation of any chemical plant. For example, a badly leaking pump conveying a hazardous liquid is working inefficiently, wasting time and materials, and may corrode unprotected parts of the pump. It might also endanger all personnel in the vicinity by risk of fire, explosion, poisoning, or burns. The design of the many types of pump available for the conveyance of liquids and gases and their particular applications is discussed in Chapter 3.

Centrifugal Pumps

The main disadvantages of centrifugal pumps are the sealing of the impeller shaft and the inability of the pump to be self-priming. The latter is only a disadvantage if the pump cannot be positioned so that its suction is flooded (see Fig. 6.1).

Where flooded suction is not possible there are two alternatives: the fitting of a non-return valve at the bottom of the suction pipe or the fitting of a self-priming unit to the volute of the pump. Both these methods ensure that the pump volute is kept full of liquid at all times; the pump will, as a consequence, be ready for

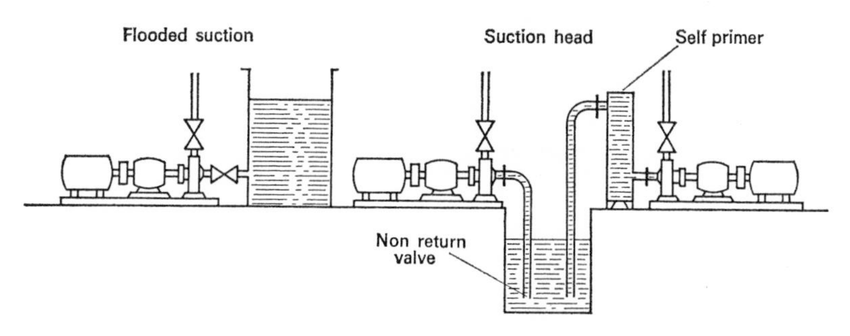

FIG. 6.1. Centrifugal pump.

immediate use. The non-return valve is not very suitable for use on slurry pumping as particles will most likely settle on the valve seat and prevent its tight closure.

The self-primer is a length of pipe or a small tank sized to hold sufficient liquor to prevent the suction pipe from draining when pumping stops. In both cases care must be taken to prevent air leaking into or liquid leaking out of the pump body or any part of the suction piping, including the self-primer if fitted. The most likely place for such a leakage to occur is at the shaft seal. All liquid must be carefully drained from such units before any attempt is made to strip the pump for maintenance work.

There is no simple solution for the sealing of the impeller shaft, although, in many cases, mechanical seals can overcome this problem. The original method of sealing the shaft is by compressing a suitable material (usually an asbestos or PTFE compound) shaped to fit the space between the shaft and the stuffing box. The pressure is maintained by tightening nuts which pull the gland-follower towards the pump casing (see Fig. 6.2).

This type of gland seal can work satisfactorily only if a small

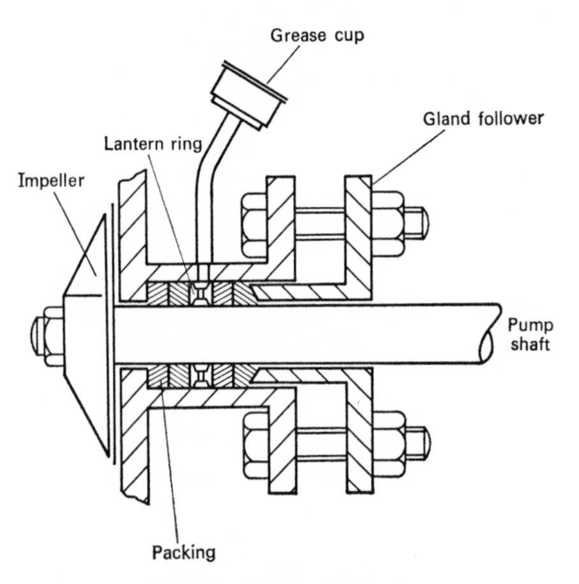

FIG. 6.2. Simple stuffing box.

amount of liquid passes through the gland, thus lubricating the packing and the shaft and preventing overheating. On some stuffing boxes nipples are fitted to permit greasing of the packing; others have lantern rings or bearing rings to permit the grease to distribute itself evenly around the shaft.

The main disadvantages of this type of seal are: (i) obtaining and maintaining the correct pressure on the packing, (ii) obtaining a really suitable packing material, (iii) the wear on the shaft shredding the packing, (iv) the regular need to repack the gland,

and (v) the undesirability of having any liquid dripping from the stuffing box.

The materials of construction of mechanical seal faces (usually carbon and ceramic) have a very wide chemical resistance and are usually unaffected by the liquor being pumped. Although seal faces must be wetted, even a very slight seepage across them is a disadvantage when slurries are being pumped, as the faces may be scratched or torn. More complicated seals, which are irrigated, are available for use when pumping slurries, but it is sometimes very difficult to obtain a suitable irrigant. With all mechanical seals the seal faces are carefully machined and lapped and they must be carefully and correctly assembled to ensure freedom from trouble. When irrigated seals are fitted the pump must not be run until the irrigant is flowing, as "dry" running may cause damage.

When operating centrifugal pumps it is particularly important that stuffing boxes drip regularly and mechanical seals should weep. It is advantageous to start a centrifugal pump against a close valve on the delivery side, and this valve should be opened slowly as soon as the pump is up to speed. In all cases, particularly in the case of pumps with a suction head, a check should be made to confirm that pumping is taking place immediately the pump has been started. If no pumping is apparent, all valves up-stream and down-stream must be checked in the open position and the pump stopped if there is still no pumping (it is probable that the pump has become air-locked). Some pumps have a small tap at the top of the volute to allow the air to be released, but if a hazardous chemical is being pumped precautions must be taken to avoid any liquid being splashed onto persons or any unprotected equipment. If air is found in the self-primer of pumps with a suction lift, then the source of the leak must be located. The stuffing box and non-return valve (if fitted) are the most likely sources.

Reciprocating Pumps

Reciprocating pumps are used mainly where high pressures are required—up to 5000 psi being attainable in a single stage. On a non-corrosive duty a pump of this type will operate trouble free, but it is dependent on tight closure of the valves for developing its pressure. It must never be started against a closed valve and must be fitted with a safety valve to limit the pressure that can be built up in the delivery line. Usually the blow-off line from the safety valve is connected back to the suction side of the pump to retain the liquid. This type of pump can also handle gases and does not require priming.

The delivery from a reciprocating pump is pulsating and this can be a disadvantage in some applications. The fitting of a receiver or balancing tank in the delivery line can markedly reduce this pulsation. The main problem in operating these pumps is associated with the valves; simple ball or dead-weight valves are usually fitted for corrosive or slurry duties if the pressure required is not high. To prevent corrosive liquors attacking the piston it is usual to operate the piston in a cylinder separated from corrosive liquor by a slug of oil or by a rubber diaphragm or tube. Many of these pumps are very reliable, can pump liquids or gases, and do not require priming.

Eccentric cylinder pumps operate by carrying fluid around the outside of the pump and forcing it out against a slide which is lifted by the eccentric drive. These pumps are often cheap, but they are dependent, like the gear pumps, on closely machined surfaces maintaining limited clearances. They are manufactured in many materials but the duty of a particular pump is limited by its materials of construction. Even slight corrosion can have a marked effect on the performance of the pump. There are many variations of eccentric pumps—Mono and Megator are only two of a wide range. Most gear or eccentric cylinder pumps are self-priming, but some require an initial wetting.

Vacuum Jets

Vacuum jets, in which water, air, or steam is forced through a carefully designed orifice so positioned in relation to the suction pipe that a partial vacuum is produced, are often used in the chemical industry. The main advantage is that corrosive vapours and gases can be readily handled as the jets are manufactured from carbon, porcelain, earthenware, or certain plastics—all chemically resistant. They are, however, often fragile and require careful handling and are susceptible to erosion, particularly when handling vapours. Where possible catch pots should be fitted in front of the jet to prevent any solids from entering. Operation of these jets is simple, but care must be taken to operate at the correct pressure of fluid to the jet; high and low pressures can cause a fall-off in the jet's efficiency. Maintenance is mainly directed to inspection for damage or corrosion of the jets.

PIPING

The material of construction of pipes differs according to the chemical being conveyed, and it is not uncommon to have as many pipes feeding a reactor as there are chemicals to be supplied. This adds considerably to the complication of pipe runs, manifolds, and valves. There is an obvious advantage in having pipes made of a material of universal chemical resistance and, although there is no material universally resistant, there are several that have a wide range of chemical resistance; for example glass or metal pipes lined with glass, Penton, polypropylene, or PTFE. These are expensive, however, and often susceptible to damage. Lined pipes need the same care and maintenance as lined vessels, but only large-diameter pipes can be inspected internally. Glass pipes require very careful assembly initially, being supported firmly but flexibly so that each pipe length is supported independantly; no attempt should be made to strain or spring them into position. All pipes should be adequately supported and allowances made for expansion where appropriate; for example a steam main should include expansion loops or bends. Even PVC piping should have expansion joints on any reasonable length that is subjected to the normal ambient temperature range.

When organic solvents are being conveyed, particularly at a high rate of flow, pipes should be adequately earthed to prevent static electricity building up; this may present a fire or explosion hazard.

Jointing

It is rarely practicable to fabricate long pipelines in one piece and for ease of manufacture, handling, storing, and erecting, pipes are usually produced in standard lengths or maximum lengths and cut to size at site. This necessitates jointing either by screwing, welding, or the use of flanges.

For many materials screwing is not possible and in many instances flanged joints are advantageous, but it is necessary to fit a *gasket* between the faces of the flanges. Rubber or asbestos are the main materials used and are often used in combination. The introduction of PTFE in the form of a sheath, fitting inside the gasket and folding along each face, has provided an almost universal gasket for chemical pipelines. Care must be taken on assembly to ensure that the sheath is lying flat and that the insert is sufficiently resilient to accommodate the surface irregularities of the flange faces.

VALVES

The design and application of the main types of valves used in the chemical industry—gate, globe, ball or plug, diaphragm, and non-return are described in Chapter 3.

Diaphragm Valves

The main advantage of the diaphragm valve is the absence of a stuffing box on the valve stem and the elimination of the consequent leaks. The diaphragms and bonnets can be removed for inspection and replacement without removing the valve from the pipeline and at the same time the valve can be inspected internally. These valves are readily lined with most materials and diaphragms and bonnets are often interchangeable between valves of different materials of construction and different lining.

Care must be taken not to overtighten when closing, particularly with chain-operated valves, as the diaphragm can be damaged if any hard material is trapped between the diaphragm and the seat.

Some diaphragm valves are straight-through design, but many diaphragm materials (particularly PTFE) are not flexible enough to allow a long movement of the diaphragm. Valves with a bridge-type seat can be tilted axially to allow a horizontal line to drain, but they should never be rodded because of the risk of damage to the diaphragm.

Non-return Valves (Check Valves)

These valves (Fig. 6.3) operate automatically and may be treated as part of the pipeline. Regular inspection, however, is needed to

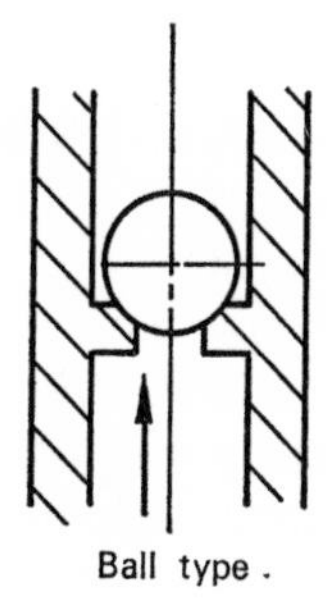

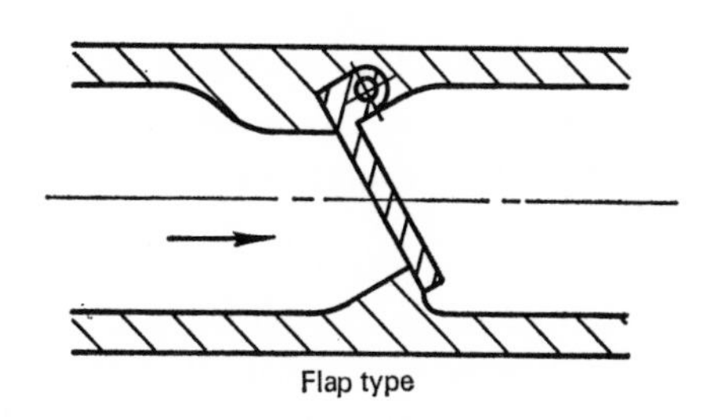

FIG. 6.3. Non-return valves.

prevent untimely failure; the seats may crack or wear as may the flap and flap pin or ball or lifting seat. The latter two types can only be used in vertical lines but the flap type will operate in horizontal as well as vertical lines. It is difficult to ensure a tight closure of these valves in slurry lines as there is a limit to the pressure that is applied to the flap or ball and even small amounts of solids on the seat will prevent a tight shut-off.

The following general points apply to all valves.

1. They should be adequately supported to prevent the line being subjected to excessive flexing when it is operated; this is most important in glass lines.
2. When a valve is in the closed position, it must be tight against

its seat, but when in the open position it should NOT be tight against the spindle stop. Hence, if the handwheel is tight, the valve is shut, but if the handwheel is loose, the valve is open.

3. If in doubt about which way to turn a valve, look on the handwheel for a directional arrow. Usually, clockwise rotation (when looking on the top of the handwheel) will close the valve.
4. Many valves have indicators showing the valve position: maintenance personnel should check that these are in good working order. The simplest and most reliable indicator is the raising stem, particularly when this is incorporated with a fixed position for the handwheel.

FANS AND DUCTING

Extractor units, although widely used for fume extraction, have the disadvantage that the motor is directly in the path of the fume and will, therefore, suffer damage if not suitably protected.

Paddle-wheel fans, designed on the same principle as a centrifugal pump, are widely used for handling corrosive fumes because the simple paddle design lends itself to easy covering with rubber, plastic, or similar protection. The motor is mounted outside the fan and does not require protection from the fumes being handled. Multi-blade, high-speed fans or blowers are more complex in "impeller" design, but give a higher throughput. These may be of rubber- or plastic-covered material, but are often built entirely of plastic. For higher pressures and larger throughputs, blowers designed on the principle of a gear pump, or exhausters designed on the lines of an eccentric pump, are used.

In practice, most fans or blowers are remotely controlled, usually situated outside a building because they are noisy in operation. In most instances their performance from an "on" or "off" aspect is evident by draught or noise. A simple U-tube manometer (see Chapter 5) is often used to measure the pressure inside and will indicate any deterioration in performance. Loss of performance is usually due to dirt build-up in the fan or ducting or on the fan blades, and its removal is part of the regular maintenance.

Frequently, ventilation systems have many suction points and they are fitted with dampers or butterfly controllers to regulate the amount of air or gas being drawn through at that point. All unused sections should be closed to concentrate the draught through those in use. It is a good practice to regulate each damper to give the minimum draw-off required at a particular time so that maximum draught can be concentrated at the points of major need. Suction points are often of flexible plastic, rubber, or metal, and care should be taken to avoid excessive flexing or bending into a tight radius. Ducting systems are designed to handle gas or air and quantities of liquids or dust or other matter inadvertently drawn into these systems often cause considerable damage. Even minor damage to a fan blade, particularly a high-speed fan, will usually set up an imbalance which may result in extensive damage to the complete unit.

Maintenance of ducting systems should include regular inspections for damage to blades, cases of fans, and also of the ducts. The direction of rotation of the fan should be checked after every electrical overhaul. All suction points should be checked for damage and evidence of liquor or dirt entry; all dampers and butterfly valves checked for ease of operation and leakage. Long vertical ducts, particularly on the delivery side of the fan, should be checked for good support; the weight of this ducting should not be taken on the fan casing.

MACHINERY

Machinery is defined as any equipment that has moving parts. It is beyond the scope of this book to discuss in detail the care of all machinery but there are some general remarks which are applicable.

Operators must not interfere with or remove any safety guards. The designed load must not be exceeded; for example the volumetic capacity, the weight to be carried, or the pressure it will withstand. The maintenance department must ensure adequate greasing and/ or oiling and regular inspection of moving parts, and check the performance of the machine. Excessive heating of machines must be reported to a supervisor.

INSTRUMENTS

For the purpose of this chapter, instruments are classified as indicators, recorders, and controllers.

The simplest instruments are those that indicate directly, for example the gauge glass (or sight glass); the former term being preferred to avoid confusion with a window which enables the inside of the vessel to be seen. Maintenance must include regular checks that valves and connecting pipes are free of obstruction and that the inside of the gauge glass is clean. Operators should keep the glass clean externally. A blockage in the top connection may give a low reading because the air above the liquid in the gauge glass will be under pressure.

Probably the most common of all instruments is the pressure gauge, indicating the pressure by means of a pointer on a circular scale (see Chapter 5). It requires practically no attention and maintenance is usually limited to regular testing. Operators must not tap or bang them as this rarely improves their accuracy but is more likely to cause damage to the instrument. On some chemical services the gauge line may get blocked or partially blocked, but the use of diaphragm gauges helps to reduce this risk. Any gauge suspected of reading incorrectly should be replaced. In many cases pressure gauges are used to indicate the contents of a vessel; the pressure, being proportional to the height of liquid above the gauge take-off point, enables the dial to be calibrated in relation to the vessel shape. The take-off point may be a direct connection to the gauge or through a diaphragm at the bottom of the tank, or it may be by means of bleeding compressed air down a dip-pipe.

In temperature indicators or thermometers the temperature is usually transmitted from a sensitive element to the dial by means of a capillary. It is important that this capillary is not kinked or bent or broken: frequent flexing will easily cause the latter. Other indicating instruments in common use in chemical factories are ammeters, measuring electric current; voltmeters for electric voltage; liquid, vapour, and gas meters; flowmeters; pH meters; weighing machines.

Recording instruments are, in principle, an adaptation of an indicating type by fitting a pen to the indicating pointer and arranging this to trace a path along a moving chart. Often the chart is a disc of paper rotating once a day or once a week, or the paper may be a long, scroll-type chart which is fed under the pen at a constant speed. The pointer must never be bent, and the cover of the instrument kept firmly closed, except when changing charts, winding the clock, or feeding ink to the pen. Most recording instruments can easily be read and are, therefore, also indicators.

Care of instruments begins when it is realized that all are delicate, that they should be kept clean, dry, not tampered with, not tapped, and not subjected to extremes of heat. The more complex, the more sensitive to maltreatment and to adjustment they become. Maintenance requires regular inspections and tests; a thorough knowledge of the instrument is required before fault finding (and adjustments) can be carried out successfully.

LIFTING EQUIPMENT

There are very few factories that do not have some form of lifting equipment; the hoist being the most common form. The hand-operated block and tackle is rarely used for production needs today, but is often the mainstay of the maintenance department for moving equipment. The Factories Act requires a high standard of inspection to be carried out regularly, but there remains the hazard of abuse or bad operation.

The safe working load (S.W.L) of any hoist must on no account be exceeded. All persons operating hoists should know the weight of the load to be lifted. All slings, chains, cradles, pallets, etc., should be kept clean—particularly in dusty or corrosive atmospheres. Chain hoists are therefore often preferred to wire hoists because inspection for corrosion is much easier. If there is a spillage over wires, chains, or hoisting hooks, they should be inspected and tested before being put into service. All unusual occurrences must be reported immediately. When a load is being lifted, precautions must be taken to ensure that no one can walk underneath the load; a person must be stationed at floor level to ensure this.

Lifts are often used for taking raw materials to a charge platform level. Again, these are inspected on a regular schedule as dictated by the Factories Act and no load in excess of the safe working load must ever be lifted. If it is marked "Goods Lift Only" then operators must use the stairs or passenger lift. Again, any chemical spillage or unusual occurence must be reported and the lift inspected before further use.

In spite of care in its operation, regular preventive maintenance, and keeping a high standard of cleanliness, chemical plant may still develop faults—but only after giving reliable service for long periods. When faults do arise, it is imperative that the operator reports them to his supervisor or the maintenance staff *immediately*. Delay may result in serious damage to equipment and risk to his safety and that of others.

When requesting the assistance of the maintenance staff, it is the responsibility of supervisors and chemical operators to ensure that safe access is provided and that faulty equipment is properly cleaned. The maintenance staff must be advised of hazardous materials contained in equipment, or if the equipment is under pressure, vacuum, or at high temperature. It should be remembered that although the fitter knows the details of the equipment, he may not necessarily know what it is being used for at a particular time. Maintenance staff must also be in possession of a certificate of authorization before entering vessels, breaking pipelines, or carrying out welding operations.

Recommended further reading:

Safety in Inspection and Maintenance of Chemical Plant. British Chemical Industry Safety Council.

CHAPTER 7

Services

GENERAL PRINCIPLES

SERVICES (sometimes called utilities) in a chemical factory are defined as the ancillary supplies and facilities required to perform the chemical processes. For example, to boil water it is necessary to supply heat—usually in the form of steam, electricity, gas, or some other flame. The following services are those usually available: steam, compressed air, vacuum, refrigeration, electricity, gases, effluent disposal, rubbish disposal, and ventilation.

It is essential to have services available in sufficient quantities and of the correct quality to meet all production requirements. Many services are required by more than one process plant and it is, therefore, often economical to produce the service required at a central point in the factory and distribute it. There may be more than one source of the same service; for example, two or more steam boilers, when it is usual to interconnect the distribution so that a more reliable supply is provided. It is clearly essential that the services must be maintained to a high standard, as their failure will affect many, or possibly all, process plants.

The principles of distribution are common to all services and usually the aim is to provide a *ring main* around the factory so that only short branches are required into each building or plant. In many instances this ring main is used as the interconnection between different sources of the same service. It is a normal practice to meter the services as they are distributed to particular plants or areas so that the quantities used by each plant are known for planning and costing purposes.

STEAM

In the past, steam has been used as a means of providing motive power, but it has now been replaced by electricity or the internal combustion engine. The main use of steam today is the supply of heat for process reactions, drying materials, general heating purposes, and for ejectors. The temperature required by the process will determine the pressure of steam required and in many processes this is not above 200 psi (Table 4). If there is a large demand for low-pressure steam and a high use of electricity, it may be economical to install a high-pressure steam boiler. This steam is used to drive a turbine generator taking the low-pressure steam exhausted from the turbine for process heating, thus providing two services concurrently.

Steam is generated in a plant on the same principle as any chemical; the raw material (water) is evaporated under pressure by the heat of combustion of oil, gas, or coal. Operators are often called boilermen: their duty is to ensure the purity of the end product, at the pressure (pounds per square inch) and in the quantity (pounds per hour) required. Usually, there is some built-in spare capacity to cover overhauls and maintenance down-time. Distribution is usually at a pressure higher than that required in any particular plant where steam is supplied through a reducing valve. This overcomes the pressure losses in distribution and enables the process pressure to be maintained. In some plants there may be a requirement for several different steam pressures and these are supplied through further reducing valves.

Safety regulations require blow-off valves and pressure gauges for each line, and it is a good practice to install these adjacent to the reducing-valve. If a gauge is added up-stream of the reducing valve, it is very easy to check that it is operating correctly. The steam user can help considerably to reduce local pressure-drops if steam valves are opened slowly, particularly when admitting steam into a cold vessel or jacket. This is most important when

TABLE 4. RELATIONSHIP BETWEEN PRESSURE AND TEMPERATURE OF SATURATED STEAM

Vacuum	Pressure	Temperature	
Inches mercury	PSI absolute	°C	°F
29	0·490	26·0	78·9
25	2·448	56·4	133·6
20	4·896	71·9	161·4
10	9·793	89·0	192·2
pressure PSI Gauge			
0	14·7	100·0	212·0
10	24·7	115·0	239·4
20	34·7	126·0	258·8
30	44·7	134·4	274·0
40	54·7	141·5	286·7
50	64·7	147·6	297·7
60	74·7	153·0	307·4
70	84·7	157·8	316·0
80	94·7	163·0	323·9
90	104·7	166·2	331·2
100	114·7	169·9	337·9
120	134·7	176·6	350·1
140	154·7	182·6	360·9
160	174·7	188·1	370·7
180	194·7	193·1	379·6
200	214·7	197·6	387·7

factory is starting up, when all vessels in all the plants are old.

Live steam points are usually avoided, but where they do exist he greatest care must be taken to avoid steam being directed at ersons; this would result in severe burns. Flexible hoses transnitting steam should be stored tidily to eliminate damage by wisting or other distortions. They must be inspected carefully efore use. Factory regulations ensure regular inspection of all team and air receivers, but nevertheless any defects must be reported immediately. Steam leaks are dangerous because the hottest steam is the invisible uncondensed vapour, not the whit mist which will appear some distance from the leak.

An essential part of every steam system is condensate removal In most applications steam is used for heating and it is the laten heat of evaporation which is given up when the steam condense that is so important. In other words, when the steam has done it work of heating there is hot condensate left which must be remove to make way for more steam. Fortunately this condensate is a pure as the treated water fed to the boiler and it is hot; therefore it may be collected and returned to the boiler. This also eliminate large volumes of flash steam being released. Part of a good stean

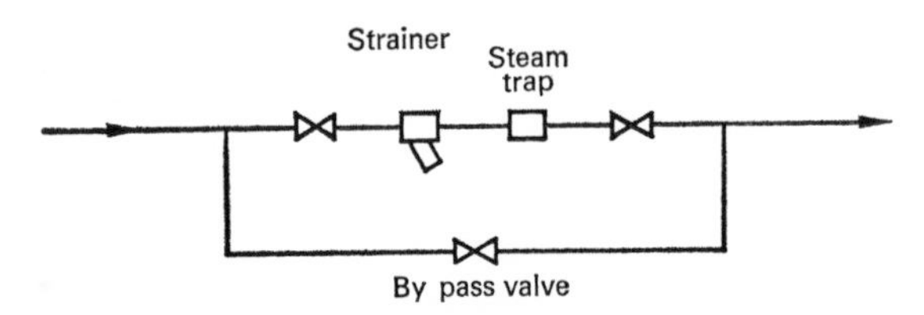

FIG. 7.1. Steam trap with strainer and by-pass valve.

distribution system is to ensure that the steam is delivered to th plant in a dry state (free of water droplets), and to ensure thi steam traps are positioned at low points on the steam main t discharge any accumulated condensate into the condensate mai without loss of steam. All available condensate is usually returne but there must be no risk of it being contaminated. For this reaso condensate is not collected from multi-service vessel jackets, tha is those that are supplied with more than one service.

Steam traps, or the strainers that protect them, are often blocke by the accumulation of rust and dirt and it is a good practice t install a by-pass valve which enables the condensate to be remove while the strainer is being cleaned; see Fig. 7.1. It is most importa to ensure that the by-pass valve is closed except while the strain or trap is being overhauled.

At the high points of a steam main air eliminators are fitte to release any accumulated cold gas or air and these are often arranged to discharge into the condensate main.

COMPRESSED AIR

Compressed air is usually distributed on a similar principle to steam, if it is required in several areas. If there is a main user, the compressor may be located adjacent to it, but it is often convenient to centralize all services in one area. As with steam, the size of the compressor and the pressure to which it raises the air will be dictated by the equipment it serves. The usual operating pressure is 100 psi and, like steam, this will be reduced where required by the use of reducing valves. The main distribution circuit may well be a ring main and it will certainly have drain points, but in this case water will be discharged from the trap, which is usually float operated.

Air which is compressed in a conventional ringed piston compressor will contain particles of oil and dirt and, when it cools, droplets of water. The bulk of these contaminants will be removed by traps, filters, dirt eliminators, air receivers, and by an after-cooler, if fitted. The resultant air is then clean enough for most applications. The air feed to air motors, which require lubricating, is passed through air lubricators. It is important for the operator to check that the transparent bowl of the air filter is clear and that the air lubricator is filled to the required oil level before using machines which have these fitted in the air supply line.

Air hoses are usually connected to the air line with self-sealing coupling, and it is important to check that these do in fact seal when the base is removed. All air leaks are a wastage which may result in a general drop in pressure if there is a significant number of them. Air leaks are usually difficult to see but are often heard (a characteristic hissing) and can be located by brushing the pipe with a soap solution. Air discharging from any open end or a leak may well contain small particles of dirt or oil which will be propelled at high velocity and are dangerous, particularly to eyes. Even clean air at high velocity could cause physical injuries. Air hoses should never be used to dust down a person's clothing, and only for blowing away dirt when there is no risk of injury to personnel in the vicinity. Air guns can now be obtained that incorporate a protective air screen for the operator.

There are many applications which call for the use of clean, dry air, for example pneumatic instruments and air masks. This need is often met by using an oil-free compressor and after-cooler or by fitting an after-cooler and air drier to a conventional compressor. This reduces the need to fit and maintain local dirt eliminators as well as reducing the number of water traps required. In some factories, both systems are in operation.

VACUUM

Vacuum is usually provided by a pump or ejector and normally supplied as a service at a *rough vacuum* (gauge reading 15–25 in. Hg) for general purposes. Long vacuum distribution systems present two major problems; they increase the incidence of leaks, which are difficult to detect, and make the attainment of high vacuum (which is really the removal of air) more difficult since this is best achieved by reducing the volume to be evacuated to a minimum. Higher vacuum (gauge reading greater than 29 in. Hg) for special purposes is provided locally as required.

In any vacuum system the operator must ensure that any unwanted lines are isolated and all vacuum points are shut off when not in use. Vacuum is easily lost by relatively small leaks which, of course, are not readily apparent as the leaks are *into* the pipe or vessel. If the required vacuum is not attained, it is best to first locate the fault by the following procedure: Isolate the vacuum pump or ejector and check that the gauge adjacent to the pump or ejector is performing correctly. If this is satisfactory, close all valves in the system and then open up each section in turn working away from the vacuum source. This will reveal the faulty section or sections and will reduce the area of search for leaks. Alternatively, the system may be pressurized and checked for leaks with soapy water as previously described.

Vacuum pumps and ejectors are often protected from contamination by means of buffer tanks, receivers, or cooled traps which will collect any dirt or liquid which may cause damage, particularly to the pumps. It is most important that these tanks are emptied regularly enough to eliminate the chance of the contaminant being drawn into the pump or ejector. When vacuum pumps are drawing the vapour of a solvent, it must be condensed in a heat exchanger before it enters the pump. Solvent drawn into a vacuum pump will remove the oil rapidly and cause the pump bearings to

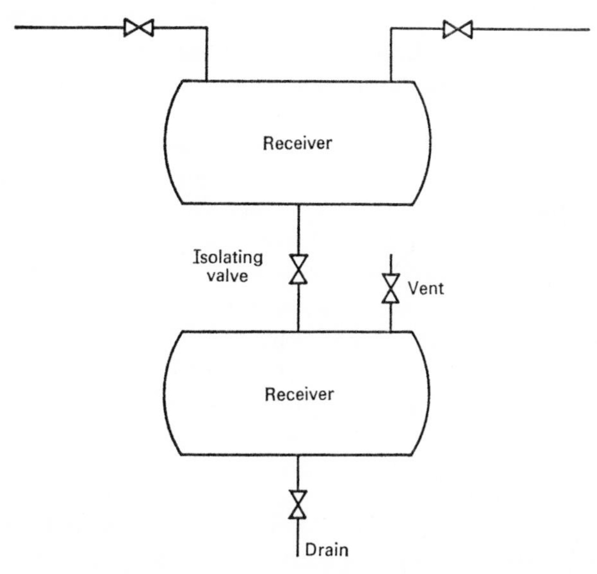

FIG. 7.2. Vacuum pump: protection from contamination.

fail. The contaminant may be removed by the installation of two tanks with an isolating valve between them. The second tank is provided with an air vent and a drain so that it can be drained without shutting off the vacuum line (Fig. 7.2).

WATER

Water supplied by the local water board, sometimes called mains water or city water, is always fit for drinking (often referred to as potable) and is normally fed directly from the mains to all drinking-water points. Only water taps labelled *Drinking Water* should be so used. Storage tanks serve as a buffer to even out surges in demand, and to provide an emergency supply. They are often placed high enough to provide sufficient pressure for the factory needs without the use of pumps. Distribution from these tanks is similar to other services, often using a ring main system.

For many chemical processes, mains water is not suitable and a high grade of *de-ionized* or distilled water is distributed from a water-treatment plant. In some cases this requirement may warrant a distribution system, and in others local treatment may be sufficient.

Water used for cooling purposes in condensers or vessel jackets is generally recirculated through a cooling tower to reduce the water consumption. In a cooling tower the water is passed down through an upward current of air which cools the water mainly by evaporation, the loss of latent heat being the major contribution to the drop in temperature. In a *natural draught tower* the increase in temperature of the air and the chimney effect are the sole driving force for the air; they are, consequently, rather tall, approximately seven-eighths of the height being above the water distribution-system. In a *forced draught tower* a fan propels the air through the tower at a predetermined rate and the tower is no higher than the top of the water distribution system. Water is distributed by flowing over a series of weirs and baffles to increase the area available for contact with the air and, in the case of a natural-draught tower, this must be much more open than in a forced-draught tower. The latter will achieve a lower water temperature and, under ideal conditions, can achieve a temperature approximately 2° C above the wet bulb temperature. When large quantities of water are involved, natural draught towers are mostly used; even though they may be as high as 300 ft and have a diameter equal to approximately two-thirds of its height.

In many factories, river water or bore-hole water is used for cooling and some form of treatment is required to prevent fouling of pipes or the precipitation of hard scale in condensers or vessel jackets. The treatment and cooling lends itself to centralization and ring main distribution. Sea water is sometimes used for cooling or quenching, but this is normally returned to source and not recirculated. Finally, fire mains are usually distributed separately supplying either sprinkler systems or a series of sealed hydrants.

The main hazard with all exposed water mains is freezing. It is important that all pipes are lagged and are maintained in good condition and suitable precautions are taken to prevent *dead legs* from freezing even when lagged. Apart from the loss of an essential service, serious damage may be caused when the ice melts. Chemical operators should remember that factories pay for each gallon of water used and every valve or tap should be closed tightly when not in use: leaks should be reported to a supervisor.

REFRIGERATION

Low-temperature coolants, at less than 0° C, include brine, which is a solution of a salt (for example, calcium chloride or sodium chloride) in water; and alcohols (for example, methyl alcohol) glycerine, and oils. The source of the low temperature is ammonia, sulphur dioxide, or carbon dioxide from a compressor. Again this lends itself to centralized production in a service area, if the demand is sufficient and, in all cases, the refrigerant is recirculated.

Although refrigerants are less likely to freeze, the pipelines are heavily lagged to prevent the absorption of heat from the atmosphere. Care should be exercised to eliminate wastage (all leaks should be reported immediately) and prevent cross-contamination with other services. This applies particularly in multi-service vessel jackets; if water is allowed to enter a refrigerant line, it may freeze and block the line, whereas a refrigerant entering a water line will contaminate water and may cause corrosion of pipelines.

ELECTRICITY

In Great Britain, electricity is generally supplied from the National Grid to most factories where it is transformed to a 440 V, 3-phase a.c. supply at 50 cycles with single-phase supply at 240 V.

The voltage drops along mains cables may be considerable, particularly if the loads are high. Transformers are often used in order to keep the various supplies within close limits. Distribution from these transformers is normally by ring main at 440 V, with single-phase 240 V supply being taken off at switchrooms in or adjacent to individual plants. Power-factor correction to offset the effect of inductive loads on the supplies may be built in to the distribution system advantageously.

Electricity is used mainly for lighting, and power for heating and motors. The former presents few problems but it should be emphasized that unnecessary burning of lights is a very common wastage. Electric motors, however, often present problems and it should be understood that in most instances the starting torque is generally high in comparison to the running torque and although an electric motor is designed to take this starting torque it is not designed to take it repeatedly in a short space of time. This means that when a motor cuts out through overload it should not be repeatedly restarted; this applies particularly when the motor is driving a stirrer or agitator in a mixture which tends to thicken. Electric motors will become very hot under these conditions and will easily burn out. The cause of the overload should be found and remedied before restarting.

Electricity is particularly dangerous because it cannot be seen; only an electrician should ever interfere with any electrical device, starter motor, connection box, or cable. Water should be kept away from all electrical equipment and any damage to electrical equipment must be put right by a skilled electrician before it is used. Special flameproof electrical equipment is provided in areas where flammable solvents are handled.

Many dust-collection and air-purification systems are electrostatic precipitors in which a wire carrying a high negative charge is placed between plates or in a pipe; the dirty particles from the air acquire a negative charge and move to the plates or pipe wall. These and any similar high-voltage equipment are always protected by means of interlocks, preventing access while the current is switched on. Interference with any such interlock would be a very serious breach of the Factories Acts.

GASES

Those gases which are provided as a service in chemical factories are usually used as fuel gases; that is their heat of combustion is utilized, for example, for heating furnaces and steam boilers. The most common gases used are town gas, natural gas, producer gas, and water gas.

Town or coal gas is that supplied from the local gas works and consists chiefly of a mixture of hydrocarbons, hydrogen, and carbon monoxide. Natural gas consists of a mixture of hydrocarbons, but natural gas from the North Sea is chiefly methane and therefore, in addition to its use as a fuel, may be employed as a starting material in the synthesis of other chemicals, for example ammonia, acetylene, and alcohols.

Producer gas, which consists chiefly of a mixture of carbon monoxide and nitrogen, is often generated in the factory either from coal in gas producers or as a by-product of a blast furnace. It is usually produced locally and the distribution kept to a minimum; if it has to be taken any distance it will require cleaning in a manner basically similar to that employed in a gas works. This cleaning involves the removal of tars, sulphur dioxide, and other gases which might foul or corrode the distribution system. The distribution system for producer gas must include explosion doors and leaks must be eliminated as carbon monoxide is a very poisonous, invisible gas with no smell. Water gas is produced by blowing steam through a bed of hot coke and consists almost entirely of carbon monoxide and hydrogen. It is used in conjuction with producer gas and is similarly distributed.

Oxygen, nitrogen, argon, and helium gases, when required in large volumes, are normally supplied from liquid in an insulated bulk storage tank and fed to a distribution main through an air- or water-heated evaporator. The pressure in such a main is normally higher than that required at the plant and the required pressure is maintaned by means of reducing valves or regulators.

EFFLUENT DISPOSAL

Effluent is a term used to describe all waste and unwanted liquors or gases resulting from chemical processes. Very few liquids may simply be run to the drain, or gases be allowed to discharge to the atmosphere. Special regulations exist for the disposal of noxious waste, and effluent systems are designed to meet these requirements. This is a very difficult task; chemical waste may be acidic and therefore corrosive, a solvent may be flammable or give rise to harmful vapours or may contain dissolved, poisonous chemicals which will pollute rivers and endanger public health. The ideal material of construction of the pipe, trough, or drain is either very expensive or difficult to obtain. Some segregation is therefore necessary; for example, strong acids could be carried in a rubber-lined main or weak acids in a lead-lined trough, the alkalis or solvents in a mild steel pipe. This segregation is also useful in treating the effluent, the acids or alkalis can be neutralized, slurries passed through some form of settling tank, and solvents separated by extraction. Where such separation exists, it is most important that the chemical operators know which waste line is provided for the particular liquid effluent they are handling.

Effluent systems must collect from every plant in the factory and a general system will usually be able to handle small quantities of any spillage, but the more it is diluted the easier it is for the system. Specific lines can, of course, accept the waste undiluted. In many factories it is not practicable to separate storm water from paved yards and roads because of possible spillages and, consequently, the effluent system must be capable of handling the surges imposed by heavy and sometimes prolonged rainfall. Where storm water is separated it is important not to pour waste down that system.

The discharge of noxious effluent gases, such as hydrogen chloride, hydrogen sulphide, and sulphur dioxide, is forbidden by the Alkali Acts. Such gases are dissolved when passed up a packed column or tower against a downward flow of an absorbing liquid. The liquid is run from the base of the column and fed into the liquid effluent system (see Chapter 4). In some cases the gas is absorbed by entrainment in water, or caustic solution circulated under pressure through an ejector. This ejector also creates a low pressure which induces the gases away from the vessel in which they are generated.

RUBBISH DISPOSAL

Rubbish disposal, like effluent disposal, presents a special problem in the chemical factories: the separation of the harmless from the hazardous material. This separation can best be done at source, that is in the plant where the rubbish is generated. It is imperative that hazardous rubbish is kept to those bins or areas assigned to it and it is not covered by a layer of harmless rubbish. The reason for separation is not only on grounds of safety, but also of economy. It is easier and cheaper to dispose of clean waste-paper or rubble than the same quantity contaminated by a hazardous chemical. The latter will require special disposal, such as incineration or neutralization.

It is also important for the disposal unit to know what wastes are being collected; some can be safely mixed, the mixing of others may cause a chemical reaction, an explosion, fire, or the liberation of noxious fumes. Usually, all rubbish is collected and stored centrally until sufficient is available for economic disposal.

VENTILATION

This service is not usually centralized, units are built into each plant or building and are often considered as part of the plant.

In many instances the chemical operator has only to switch the fan or fans on or off as required. Care should be taken to avoid blocking the open ends of the ducts and to use the control dampers, where supplied, to give the minimum ventilation required. This enables more draught to be concentrated in areas which may require it. It is pointless to have all dampers fully open all the time.

Most services are supplied on a factory-wide basis, and are therefore to be shared by many plants or departments; the operator must not be "greedy", the systems are designed to meet all needs—

not every excess demand. Services are on "tap" but it must be remembered that they do cost money to operate and if the "tap" is left turned on unnecessarily there is a wastage, probably coupled with a shortage somewhere else.

CHAPTER 8

Safety in Chemical Plant Operation

THE new chemical operator may feel that working with chemicals is a hazardous occupation—he learns, for example, that they may be corrosive, poisonous, irritating, dermatitic, flammable, or explosive. In fact, the safety record of the chemical industry, although it could and must be improved, compares favourably with most other industries.

An extract from the statistics for industrial accidents in 1965, issued by Ro.S.P.A., is given in Table 5. The B.C.I.S.C.* have reported an accident frequency rate for its member companies of 1.59 for the period 1964–7 and in its report *Safe and Sound* (1969) calls for a target of 0·5!

The following statement is often made since it defines the basic approach to the problem of safety.

> *Chemicals in any form can be safely stored, handled, or used if the physical, chemical, and hazardous properties are fully understood and the necessary precautions, including the use of proper safeguards and personal protective equipment, are observed.*

In this chapter it is intended to show that, whatever safety

* British Chemical Industry Safety Council.

TABLE 5

Business	Accident frequency rate*
Timber, furniture, etc.	3·9
Shipbuilding and marine engineering	3·8
Mining and quarrying	3·7
Gas, electricity, water	3·3
Construction	3·1
Leather, leather goods, and fur	2·9
Paper, printing, and publishing	2·4
Food, drink, and tobacco	2·3
Textiles	2·2
Engineering and electrical goods	1·9
Chemical and allied industries	**1·8**
Distributive trades	1·6
Vehicles	1·2
Clothing and footwear	1·1

$$* \text{ Frequency rate} = \frac{\text{Number of lost-time accidents} \times 100{,}000}{\text{Man-hours worked}}.$$

provisions are made in equipment design or operational techniques, the ultimate responsibility for safety rests on the individual. The chemical operator must constantly discipline himself to employ safe working practices and be alert to potential danger.

WHY DO ACCIDENTS OCCUR?

A fully trained chemical operative will handle all chemicals with confidence in his personal safety, but he will never allow himself to become over confident, to take chances or short cuts. He knows that all chemicals must be treated with respect and, even when handling comparatively harmless materials, he will not deviate from proper handling methods since unsafe practices can easily become a habit, and bad habits are the hardest to break.

Unfortunately, there are some chemical operatives who are misguided enough to think that their workmates would consider it unmanly, or even cowardly, to wear the full personal protective equipment for a hazardous operation. For the same reasons they may not report unsafe acts and conditions to their supervisor. It is regrettable, but often true, that only *after* a serious accident has occurred do supervisors receive a host of complaints about unsafe conditions and suggestions for improvements. Like the car driver who only becomes safety conscious after being involved in, or having witnessed, a serious road accident. There are also those who consider themselves immune from any danger and set about their work in the manner of "it could never happen to me". This complacency can be fatal.

Dressing up in full protective clothing is often considered time-wasting. In fact it takes less than 3 minutes to dress in a plastic suit, protective boots, safety hat, face visor, and gloves, which provide complete head to toe protection for many operations which might otherwise be hazardous. Less than 3 minutes, measured against hours, weeks, or months of pain and loss of earnings.

Time spent on safety is time well spent!

Most accidents arise through lack of training or lack of concentration—the proper way to do a job is the safest way and this requires thinking before acting.

When in doubt, always ask your supervisor.

BASIC SAFETY RULES

Special safety rules associated with specific chemical plant operations are mentioned in other chapters, but there are a few basic safety rules which must be observed when working in all chemical manufacturing areas:

NO smoking or eating.
NO drinking of alcoholic liquor.
NO drinking, except from water fountains.
Safety instructions and danger signs must be observed.

Wear all personal protective equipment provided and replace damaged items immediately.

Do not enter processing areas other than your own place of work.

Do not tamper with faulty equipment or electrical fittings—report such faults to your supervisor.

Get immediate medical attention for injuries, however minor they appear to be. Report all accidents to your supervisor.

Do not lose your temper, or indulge in "horse-play".

Know the location of fire alarms, extinguishers, and escape routes.

When in doubt—ask your supervisor.

DETECTION OF POSSIBLE HAZARDS

The skilled chemical operator uses all his senses to detect and recognize the symptoms of a potential safety hazard. He will get himself—and others—clear of the source of danger and report his observations to his supervisor immediately. Most dangers can be avoided if readily recognized and the right corrective action taken promptly.

Some of the ways in which the signs of possible hazards may be recognized are as follows:

Seeing

Look for warning signs, tags, and safety instructions. Smoke, fumes, or sparks, particularly from electrical equipment, are symptoms of overheating and constitute a fire hazard. A blue flash may indicate a discharge of static electricity. Observe gauges for sudden increases in temperature, rapid pressure changes, rise or fall in liquid levels in tanks and gauge glasses. Look for cracks or other visible damage to equipment, particularly glassware and glass linings and leaks at valves, gaskets, or seals. Keep watch for physical changes of chemicals, such as a change in colour, lumpiness, change in crystal form, bubbling, or frothing. Do not poke your head into any vessel to see it is clean. Look out for reddening of the skin or rashes; they may indicate contact with an irritating, corrosive, or dermatitic chemical.

Hearing

Whistling or hissing noises may be a warning of the escape of gas or steam under pressure. Banging, rattling, grinding, or whining sounds usually indicate faults in moving parts of machinery. Hammering or knocking noises in pipes usually indicate rapid pressure changes, surging, or a liquid–vapour mixture in the line. A dripping or splashing noise may lead to the detection of leaking vessels, pump seals, valves, or gaskets. The sound of shattering or cracking of glass is associated with fracturing of glass pipeline, glass vessels, or fittings caused by excessive pressure, being struck or being subjected to a sudden change in temperature (thermal shock). A cracking noise may indicate the discharge of static electricity which constitutes a fire hazard.

Feeling

Unusual vibrations indicate pressure changes or uneven running, for example misaligned moving parts or a basket of a centrifuge loaded unevenly. Excessive heat radiating from reaction vessels may be a warning of an exothermic reaction; in the case of electrical equipment or machinery it may indicate an unusual overload due, perhaps, to loss of lubricant. Burning, irritation, or itching of the skin indicates contact with corrosives, irritants, or dermatitic chemicals. Dryness of the lips or skin may indicate contact with organic solvents. Smarting, irritating, watering, or itching of the eyes may be caused by contact with certain dusts, or lachrymatory vapours. Difficulty in breathing, choking, giddiness, or the feeling of weakness at the knees may be the effects of harmful gases, dusts, or vapours.

Smelling

The chemical operator learns to identify certain chemicals by their distinctive smell and he may be forewarned of possible danger. It is often difficult to describe a particular smell, and the sense of smell varies considerably from person to person; experience is the only reliable guide. Some chemicals may irritate the mucous membranes; some are sweet and sickly (such as nitrous oxide); others are pungent and suffocating and may cause gasping (such as ammonia). A pungent, irritating smell is characteristic of chlorine or bromine: a sweet, scented smell may indicate the organic chemicals called esters (such as amyl acetate). Organic compounds of benzene give a characteristic *aromatic* smell which is easily recognizable. There are also many chemicals which give smells identifiable with common day-to-day smells. For example, hydrogen sulphide (rotten eggs) or carbon disulphide (decaying vegetables). Even the absence of a smell may, in some cases, cause one to suspect the identity of a material. In all cases, smell should only be used to detect danger from general atmospheric polution. NEVER deliberately sniff at a chemical, under any circumstances.

Tasting

Chemicals must NEVER be tasted as this is a very dangerous practice. However, in the event of accidental contact with the lips or mouth, an acute sense of taste may help the person affected to take the appropriate action quickly by washing the mouth with copious quantites of cold water for at least 15 minutes. For example, the sour taste of acids, the sweet, sickly taste of nitrous oxide, or even the loss of taste caused, for example by phenol, may be recognized readily.

HOUSEKEEPING

Generally, the efficiency of any industrial plant can be judged by the standard of its cleanliness and tidiness. In the chemical industry in particular, poor housekeeping is a challenge to safety. Thus, housekeeping deserves special mention in this chapter.

A chemical operator who allows his place of work to become dirty and untidy is a danger to himself and others. If he allows exits to become blocked, or litter in gangways, he may cause a person to trip and fall; if he fails to clean up a liquid spillage, it may cause a fire or burns, or a person to slip or fall. Objects must not be placed under safety showers or block access to fire-fighting equipment.

Some chemical operators feel that cleaning and tidying up should be left to others, but housekeeping is an integral part of chemical operations. Operations must be carried out in a manner that the need to clean up after a job is kept to a minimum. It should be remembered that even the most senior chemist working in a laboratory cleans and tidies his own bench.

PERSONAL PROTECTIVE CLOTHING AND EQUIPMENT

A wide range of clothing and equipment for personal protection in a variety of materials is generally available, and the chemical operative should seek the advice of his supervisor as to the correct type to use for each task he may be required to perform. It is important to remember that a chemical operator is entitled to refuse to do a job if the appropriate protective clothing or equipment is not available; on the other hand, it is the responsibility of his supervisor to ensure that it is available and worn where necessary.

Safety Hats

Soft caps of plastic or leather give protection against chemical splashes, especially when working with overhead pipes, tanks, heat exchangers, and other equipment, which may leak. Reinforced hats of metal, laminated plastics, or other materials resistant to impact from falling objects, should be worn when overhead work is performed (a properly fitting hat gives maximum protection).

Dust Masks

Many types are now available, all giving protection against the inhalation of harmful dusts. Special absorbent pads are covered by a perforated metal disc or fitted to a moulded rubber facepiece: the pads should be changed regularly and the facepiece cleaned after use. It is most important to remember that dust masks offer no protection against gases.

Air Masks

The most common air masks are of the "army gas-mask" type: a moulded rubber front with two eye ports is held to face by adjustable elastic straps fitting around the head. Air is drawn through a flexible hose and a non-return valve, in front of the mouth, from a canister strapped to the body, and is expelled between the cheeks and the rubber sides of the mask. The canister contains a suitable, absorbent material and it is therefore of VITAL importance to distinguish the correct canister, for the conditions to be entered. Check with your supervisor. There is a limited life before the absorbent material is spent and it is susceptible to deterioration, so immediately it is suspected that the mask is leaking, leave the contaminated area. To obtain protection over longer periods, there are similar masks which are supplied by air from a compressed air line or cylinder, or from a hand or electrically driven air blower (positioned in the "fresh air"). The canister-type mask gives the wearer the most freedom of movement and those connected to the blower the least. There is a danger of the trailing air lines becoming jammed or cut or of tripping up the wearer.

Safety Footwear

Industrial safety shoes and boots, with steel toe-caps, which are of good appearance and comfortable, are usually supplied free or at reduced cost. Ordinary shoes are most unsuitable as they offer little resistance to corrosive chemicals or to falling objects. Sparks from nailed boots are a source of danger. Rubber boots are watertight and resistant to most corrosive chemicals, but may be attacked by many organic solvents. Your supervisor will advise you of the correct footwear for your particular job.

Eye Protectors

Eye protection against chemical splashes is provided by safety spectacles, goggles, face shields, visors, or masks. Spectacles with reinforced glass give the least protection, but have the advantage that they may be worn for long periods without discomfort. Some spectacles are fitted with plastic side-shields for added protection. Perspex goggles, with a surrounding shield of plastic fitting tightly to the face around the eyes, are particularly useful, not only against chemical splashes, but also in the presence of dust or gases which affect the eyes. Face shields or visors give full protection against splashes or sparks, and although they tend to "steam up" under certain conditions, they may be treated with "demisters". Remember, eyes can NEVER be replaced.

Gloves

Many chemicals are corrosive to the skin or may be poisonous when absorbed through the skin. Gloves are available in rubber, plastic, leather, and asbestos—each suitable for a specific type of chemical operation. Leather gloves are worn for operations which might otherwise be abrasive to the skin and asbestos gloves may be used for handling hot materials. Rubber gloves are resistant to acids and alkalis but are not suitable for many solvents. Plastic (PVC) gloves offer good resistance to attack by most chemicals, but tend to be stiff and cause the hands to sweat.

Whatever type of glove is worn, it is essential to keep them well maintained. They should be carefully inspected and washed after use. Even a pin-hole leak cannot be ignored as this may permit the intrusion of a sufficient quantity of corrosive or poisonous material to cause injury. Rubber or plastic gloves may be filled with water to check for leaks.

Safety Suits

Safety suits are available in a range of materials, for example plastic, asbestos, or rubber. Designed to provide protection for the whole body, they have overlapping edges, press-studded or zip-fastened jacket and one-piece trousers fitting over calf-length rubber boots. The helmet (or hood) is supplied with air from a breathing apparatus, a hand or mechanically operated pump, or a cylinder of compressed air strapped to the back.

Barrier Creams

Special creams or lotions which protect the hands by preventing direct contact of chemicals with the skin may be effective in avoiding dermatitis or other skin infections. There are two main types: water repellent and oil repellent. Preparations are now available which claim to be both water and oil repellent. Advice should be sought on their use, since some types, under certain conditions, may actually *increase* the risk of skin infection. There is a limit to the quantity of harmful material a layer of barrier cream can repel, and operators might be tempted to handle material indiscriminately thinking they are fully protected. In all cases, regular washing of the hands with plenty of soap and water is a defence against skin injury.

Safety rules associated with specific chemical operations are mentioned elsewhere, but the general safety aspects of working with systems under high pressure or vacuum are given special mention below.

PRESSURE SYSTEMS

Equipment that operates under pressure is specially designed to withstand that pressure and is fitted with a safety or pressure release valve or bursting disc to prevent excessive pressure developing inside it. The safe working pressure (S.W.P.) should be clearly marked on the equipment: and it is important that this pressure is not exceeded. The gauge pressure must be observed regularly.

Before pressurizing a system, all valve settings must be checked and all manways, flanges, covers, etc., must be bolted securely using *all* the clamps or bolts provided.

When opening up a pressure system, ensure that the pressure has been released completely through the proper vent. This is essential to avoid the risk of being sprayed with its chemical contents, releasing flammable or noxious vapours or being injured by the blast. Jets of compressed air may contain dust or grit which can cause injury when moving at high speed; the air itself has been known to kill people by injection into the blood stream.

VACUUM SYSTEMS

Systems from which air has been evacuated are subject to mechanical strain through external atmospheric pressure and must be handled with care. Vacuum must be applied *slowly*. The effect of a vessel collapsing (sometimes called an implosion) is similar to that of an explosion as its contents may be flung outwards. Reducing the internal pressure lowers the boiling point of liquids which may be contained in the vessel causing them to "bump" or bubble violently and be carried over through the outlet of the vessel. Vacuum must also be released *slowly*. A sudden rush of air into a system places it under mechanical strain and, in certain circumstances, may cause fire or explosion of its contents. For the latter reason vacuum is often released by admitting an inert gas, such as nitrogen or argon, instead of air.

Rapid advances are being made towards a higher degree of safety in the chemical industry—in working conditions, equipment design, operational techniques, safety devices, protective clothing and equipment, and our knowledge of specific chemical hazards. These advances, however, are still ultimately dependant on the individual chemical operator attaining a high degree of skill and knowledge of safe techniques and safety rules.

The safe way is always the best way!

Recommended further reading (published by B.C.I.S.C.):

Protection of the Eyes.
Safety Training—A Guide for the Chemical Industry.
A Guide to Fire Prevention in the Chemical Industry.
Safe and Sound.

Test Questions

1. Name *two* metallic and *two* non-metallic materials. Compare and contrast their physical properties and give one example of their use as materials of construction of chemical plant.

2. Write brief notes on the composition and manufacture of glass. Describe its applications in the construction of chemical plant, indicating its advantages and disadvantages as a material of construction.

3. Which of the following materials are alloys: aluminium, nickel, stainless steel, bronze, titanium, brass, lead, tantalum? State the composition of *one* of them and give examples of its use as a material of construction of chemical plant.

4. Describe *two* types of conveyor designed to handle solid materials in chemical factories. What are the factors which influence the choice of conveyor for a particular purpose?

5. Describe *two* methods, other than the use of pumps, used to convey liquid chemicals in the factory, and indicate any hazards which may be involved. State the advantages and disadvantages of each method.

6. Write a short essay on the construction, use, and care of gas cylinders.

7. Review the general methods for sampling solids and liquids.

8. Draw a simple flow diagram (or flowsheet) for any processing plant. Where possible, indicate the materials of construction of each item of equipment and the route of process materials through this equipment.

9. What is the purpose of packing material in towers or columns? Name and describe, with the aid of a sketch, *two* types of packing.

10. Describe, with the aid of a diagram, any chemical reactor which is used in your factory. Indicate its material of construction, the type of agitator, and the means by which raw materials are fed to it.

11. Draw a labelled diagram of a simple filter. Name *three* common types of filter and, in each case, explain the nature of the force employed to effect the separation.

12. Draw a labelled diagram of *two* of the following: (a) diaphragm valve, (b) gate valve, (c) non-return (check) valve.

13. Explain any *four* of the following terms: reflux ratio, unit load, pH, mother liquor, seeding, pre-coat, raffinate, calibration.

14. Define and describe *one* of the following unit operations: calcination, distillation, evaporation, solvent extraction (liquid–liquid). Give an example of the application of any basic unit operation.

15. State the units of measurement of the following: temperature, pressure, volume, electrical current. Describe, with a diagram, the instrument you would use to measure one of them.

16. State Ohm's law. Draw a labelled diagram of a moving iron instrument for the measurement of electric current.

17. With the aid of a labelled diagram, describe the operation and use of a Bourdon gauge. What is the relationship between atmospheric pressure and gauge pressure?

18. What is meant by calibration? Describe how you would calibrate a gauge (or sight) glass or a dipstick for a particular vessel.

19. Write brief notes on the care and maintenance of (a) glass-lined equipment, (b) pumps.

20. Discuss the provision and use of *two* of the following services (utilities) in chemical factories: steam, brine, vacuum, electricity, fuel gases.

21. Write a short essay on the uses of compressed air in a chemical factory, with particular reference to safety factors.

22. Describe any accident which has occurred (or *could* occur) in your own factory. Discuss the ways in which, in your opinion, it could have been (or *could* be) avoided.

23. Draw up a list of *ten* basic safety rules which might be applied in any chemical factory. State which you think is most important and why.

24. Discuss the potential hazards associated with the operation of pressure and vacuum systems, and the special safety precautions required.

25. "*Poor housekeeping in a chemical plant is a challenge to safety.*" Discuss this statement with particular reference to any plant in your own factory.

Index

INDEX

INDEX